AF581788

New Trends on the Combustion Processes in Spark Ignition Engines

New Trends on the Combustion Processes in Spark Ignition Engines

Editor

Jorge Martins

MDPI • Basel • Beijing • Wuhan • Barcelona • Belgrade • Manchester • Tokyo • Cluj • Tianjin

Editor
Jorge Martins
Mechanical Engineering
Department
Universidade do Minho
Guimaraes
Portugal

Editorial Office
MDPI
St. Alban-Anlage 66
4052 Basel, Switzerland

This is a reprint of articles from the Special Issue published online in the open access journal *Energies* (ISSN 1996-1073) (available at: www.mdpi.com/journal/energies/special_issues/spark_ignition_engines).

For citation purposes, cite each article independently as indicated on the article page online and as indicated below:

LastName, A.A.; LastName, B.B.; LastName, C.C. Article Title. *Journal Name* **Year,** *Volume Number,* Page Range.

ISBN 978-3-0365-1448-2 (Hbk)
ISBN 978-3-0365-1447-5 (PDF)

Contents

About the Editor

Jorge Martins

Jorge Martins is a professor at the Department of Mechanical Engineering of the Universidade do Minho, Guimaraes, Portugal, where he lectures on the subjects of thermodynamics, internal combustion engines and alternative energies. He is the head of the Laboratory of Thermal Engines, where various national and international projects are being developed.

He is the author of three books, *Internal Combustion Engines, Electric Cars* and *Accident Reconstruction* (all in Portuguese); four patents; and over 100 journal and conference publications.

Preface to "New Trends on the Combustion Processes in Spark Ignition Engines"

Throughout the world, governments are anticipating the elimination of the use of conventional fuels in internal combustion engines starting in 2030 or 2035. Although they refer to the of combustion engines, what is actually needed is the reduction or elimination of fossil CO_2 production and pollutant emissions, which is not only an engine issue but also primarily a fuel matter. Such reductions in CO_2 and pollutants may be achieved by using and adapting IC engines to various types of biofuels (to reduce fossil CO_2) and by enhancing engine design and their combustion.

This Special Issue on Trends on the Combustion Processes in Spark Ignition Enginesfocuses on some of the measures and processes required to minimise these problems in order to reduce fossil CO_2 and pollutant emissions through the use of alternative fuels, by better understanding combustion processes and by developing new methods and technologies aimed at improving engine efficiency.

Jorge Martins
Editor

Article

Performance and Emissions of a Spark Ignition Engine Operated with Gasoline Supplemented with Pyrogasoline and Ethanol

Luís Durão [1,2], Joaquim Costa [3], Tiago Arantes [3], F. P. Brito [3], Jorge Martins [3] and Margarida Gonçalves [1,2,*]

1 MEtRICs—Science and Technology of Biomass Department, Faculty of Science and Technology, Universidade NOVA de Lisboa, 2825-149 Caparica, Portugal; luisdurao@gmail.com

2 VALORIZA—Research Centre for Endogenous Resource Valorization, Instituto Politécnico de Portalegre, 7300-110 Portalegre, Portugal

3 MEtRICs—Mechanical Engineering Department, Universidade do Minho, 4800-058 Guimarães, Portugal; joaquimdacosta77@yahoo.com (J.C.); a59846@alumni.uminho.pt (T.A.); francisco@dem.uminho.pt (F.P.B.); jmartins@dem.uminho.pt (J.M.)

* Correspondence: mmpg@fct.unl.pt

Received: 13 August 2020; Accepted: 3 September 2020; Published: 8 September 2020

Abstract: The partial replacement of fossil fuels by biofuels contributes to a reduction of CO_2 emissions, alleviating the greenhouse effect and climate changes. Furthermore, fuels produced from waste biomass materials have no impact on agricultural land use and reduce deposition of such wastes in landfills. In this paper we evaluate the addition of pyrolysis biogasoline (pyrogasoline) as an additive for fossil gasoline. Pyrogasoline was produced from used cooking oils unfit to produce biodiesel. This study was based on a set of engine tests using binary and ternary mixtures of gasoline with 0, 2.5, and 5% pyrogasoline and ethanol. The use of ternary blends of gasoline and two different biofuels was tested with the purpose of achieving optimal combustion conditions and lower emissions, taking advantage of synergistic effects due to the different properties and chemical compositions of those biofuels. The tests were performed on a spark-ignition engine, operated at full load (100% throttle, or WOT—wide open throttle) between 2000 and 6000 rpm, while recording engine performance and exhaust gases pollutants data. Binary mixtures with pyrogasoline did not improve or worsen the engine's performance, but the ternary mixtures (gasoline + pyrogasoline + ethanol) positively improved the engine's performance with torque gains between 0.8 and 3.1% compared to gasoline. All fuels presented CO and unburned hydrocarbons emissions below those produced by this type of engine operated under normal (fossil) gasoline. On the other hand, NO_x emissions from oxygenated fuels had contradictory behaviour compared to gasoline. If we consider the gains achieved by the torque with the ternary mixtures and reductions in polluting emissions obtained by mixtures with pyrogasoline, a future for this fuel can be foreseen as a partial replacement of fossil gasoline.

Keywords: lipid bio-oils; pyrogasoline; performance; exhaust emissions and spark-ignition engine

1. Introduction

Throughout the world, cars are essential in everyday life and related industries, and are a source of jobs and economic growth. Internal combustion engines will continue to be the main means of propelling cars. In the current phase of transition to sustainable mobility, good energetic performance and minimum emissions are critical. In 2050 more than half of the passenger vehicles sold are expected to continue to be equipped with spark-ignition (SI) engines (gasoline, compressed natural gas/liquid petroleum gases (CNG/LPG), and gasoline hybrids) [1]. CO_2 emissions (the main greenhouse gas (GHG)), which are ubiquitous in combustion engines, are related to fossil fuel consumption [2].

Concerning CO_2 emissions, biofuels tend to be neutral and can be used in mixed or pure form in engines. The incorporation of biofuels will allow vehicles to achieve significant reductions in emissions of some pollutants and GHG [3], but it is not feasible to replace all fossil fuel consumption with them [4], with the exception of in Brazil [5]. However, the European Commission [6] introduced the term indirect land use change (ILUC) to account for the consequences on the land use (sustainability) in the production of biofuels. Therefore, the use of waste biomass materials as feedstocks for bioenergy and biofuels is promoted in the European Union as a tool for reducing the deposition of such materials in landfills and of increasing the production advanced biofuels. The Annex IX of the Directive (EU) 2018/2001 lists a group of waste biomass materials (including used cooking oils) whose contribution for the renewable energy shares may be considered to be twice their energy content [7]. On the other hand, changes in consumption patterns have also led to massive waste generation [8]. Many wastes have no economic value and cause environmental problems because of their disposal; that is the case with lipid wastes, such as used cooking oils or animal fats unfit for biodiesel production, that often are disposed of without any kind of material or energetic valorisation [9]. Pyrolysis is a thermochemical process that allows conversion of low-quality lipid wastes into carbon-rich bio-oils that may then be distilled or upgraded to yield advanced biofuels appropriate for replacement of gasoline or diesel.

During pyrolysis, triglycerides (which are the main components of the lipid wastes) undergo cracking and condensation reactions to yield the pyrolysis bio-oil, a mixture of compounds consisting mainly of hydrocarbons, oxygenated compounds, some gaseous products (CO, CO_2, and water), and a carbonaceous residue containing the mineral components of the wastes and some high molecular weight organic products [10].

The process temperature and residence time influence the pyrolysis reactions, affecting product yield and composition, but further selectivity may be achieved using modified atmospheres and different types of catalysts [11,12]. It was observed that zeolites were suitable catalysts for the cracking processes of vegetable oils to obtain compounds with boiling points in the range of gasoline [13].

The use of gasoline with added alcohol is a practical way of improving the octane number (ON) of conventional gasoline [14], often used in motor sports. The latent heat of vaporization of ethanol is 2.5 times higher than that of gasoline, allowing an increase in the volumetric efficiency of the engine [15] by lowering its intake mixture temperature. Ethanol has a high octane number and contains oxygen, therefore, its mixture with gasoline reduces the tendency to "knock" and promotes the reduction of emissions of some exhaust gases [16].

Oztop et al. [17] evaluated the performance and emissions of exhaust gases of an SI engine fuelled with a mixture of gasoline and pyrolysis distillates from tire wastes. The authors concluded that this new fuel could partially replace gasoline blends by up to 60% without significant changes in engine performance and exhaust emissions.

Suiuay et al. [18] evaluated the performance and emissions of exhaust gases of an SI engine fuelled with a mixture of gasoline and pyrolysis bio-oil distillates from the hard resin of Yang (gasoline-like fuel (GLF)). The authors concluded that the GLF showed better results than gasoline for torque, brake thermal efficiency, and brake specific fuel consumption; GLF had lower emissions of CO and unburned hydrocarbons (UHC) and higher emissions of NO_x.

The use of ternary mixtures of gasoline and two different biofuels has been tested by different authors with the purpose of achieving optimal combustion conditions by combining fuels with different chemical composition and fuel properties [19,20].

Kareddula et al. [19,20] published two papers evaluating the performance and emissions of exhaust gases of an SI engine with pyrolysis bio-oils of non-distilled plastic waste (PPO) and distilled PPO (DPPO). The first test was performed with 15% PPO and showed a reduction in thermal efficiency and a substantial increase in NO_x emissions [19]. Next, they tested gasoline with 15% PPO and 5% ethanol and observed that the engine's performance was improved compared to gasoline and gasoline with PPO; CO and NO_x emissions were significantly reduced while UHC emissions increased [19]. For the second study [20], they evaluated gasoline blends with different DPPO levels. The authors verified that the mixture with 50% DPPO provided the maximum performance (power and thermal efficiency) when compared to gasoline. Mixtures with lower levels of DPPO produced a decrease in CO_2 and UHC emissions and an increase in CO and NO_x [20].

The incorporation of a third component with different characteristics may achieve the attenuation of unwanted characteristics of a given fuel blend. That was the case for the addition of ethanol to the mixture of gasoline + bio-oil from plastic pyrolysis (PPO) that improved the engine's performance, with NO_x emissions being marginally controlled. Gasoline and PPO are perfectly miscible, but the higher density and viscosity of PPO may negatively impact engine performance of the fuel blend, thus, ethanol addition may attenuate these properties and increase oxygen local availability, improving mass transfer and combustion efficiency [19].

Bio-oil or bio-oil distillates derived from materials originating from oil (plastics and tires) were used for these three studies [17,19,20].

There are very few studies that evaluate the use of distillates from bio-oils mixed with gasoline in engines. Reported tests with ternary mixtures of gasoline with bio-oil distillates and ethanol in small percentages are even rarer. We did not identify any work whose analysis relied on the use of light distillates of lipid bio-oils in engines.

The aim of this study was to evaluate the feasibility of adding small quantities of pyrogasoline (a lipid-derived biofuel produced by pyrolysis and distillation), as an additive for gasoline. To achieve this goal, different binary and ternary mixtures of gasoline, pyrogasoline, and ethanol were used as fuels in an SI engine, and the impact of fuel composition in engine performance (torque, energy consumption, and efficiency) and pollutant emissions of exhaust gases (CO, UHC, NO_x) was evaluated.

2. Experimental Procedures

2.1. Pyrolysis Bio-Oil Distillation

Liquid pyrolysis products, obtained from various lipid raw materials unfit for the production of biodiesel (used cooking oil, high acid poultry oil, palm oil, high acidity olive oil, and olive husk oil) and under different operating conditions (temperature, pressure, residence time, and different atmospheres), were combined and distilled together to obtain a volume of distillate suitable for engine combustion tests.

The pyrolysis liquids were first decanted to isolate the aqueous fraction from the organic, followed by distillation to separate volatile and non-volatile bio-oil components (Figure 1). Distillation led to two liquid fractions: a more volatile fraction (F1) collected between room temperature and 195 °C and a less volatile one (F2) recovered between 195 °C and the final distillation temperature (around 250 °C).

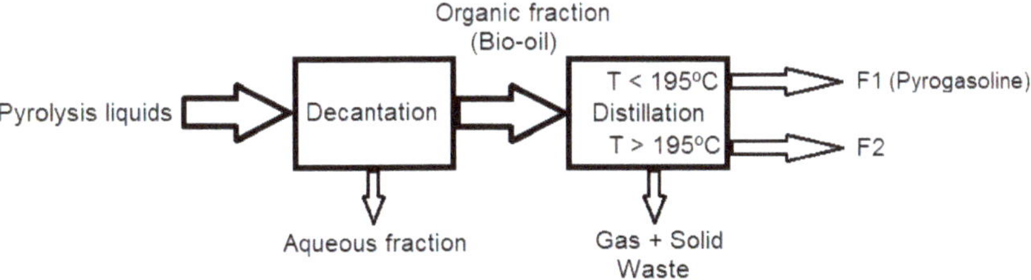

Figure 1. Distillation yields.

The crude bio-oil and light fraction of distillation (F1, pyrogasoline) presented high higher heating values (HHVs), respectively 41.3 and 42.2 MJ kg^{-1}, which gives them excellent fuel properties. Only fraction F1 (pyrogasoline) was used in this work.

2.2. Fuels for the Tests

The engine tests were performed with six fuels: five were mixtures obtained from gasoline with pyrogasoline and/or ethanol; the sixth was gasoline used as a reference fuel. Fuels are identified in Table 1.

Table 1. Identification of the fuels used in the engine tests (%wt.).

Fuel Code	Gasoline (G)	Pyrogasoline (PG)	Ethanol (E)
G100	100		
G97 + PG3	97	3	
G97 + PG1.5 + E1.5	97	1.5	1.5
G95 + PG5	95	5	
G95 + E5	95		5
G95 + PG2.5 + E2.5	95	2.5	2.5

Gasoline (RON 95) and ethanol (96% *v/v*) were purchased on the market.

The density, elemental composition, higher heating value (HHV), and functional groups were measured for each fuel. The density was determined gravimetrically. The elemental composition (C, H, N, and S) was determined using a Flash EA 112 CHNS analyser (Thermo Fisher Scientific, Waltham, MA USA), and the oxygen concentration was obtained by the difference. To determine the HHV, a model *C200* calorimeter (IKA, Staufen, Germany) was used and the liquids were weighted and analysed using model *C9* gel capsules from the same manufacturer. HHV values were obtained indirectly using Equation (1):

$$m_{total} * HHV_{total} = m_{gel} * HHV_{gel} + m_{fuel} * HHV_{fuel} \tag{1}$$

The lower heating value (LHV) was estimated using Equation (2) (ASTM D249, 2002):

$$LHV\left(MJ.kg^{-1}\right) = HHV - 0.2122 \times H(\%) \tag{2}$$

To identify the functional groups, a Fourier Transform Infrared (FT-IR) spectrophotometer from Nicolet iS10, (Thermo Scientific, Waltham, MA USA) was used; to collect the spectra and identify the main bands, the TS OMNIC program was used.

2.3. Experimental Setup

The tests were carried out on a 1.6 L, L4, 4-stroke gasoline engine (Figure 2) that maintained most of the original characteristics, except the intake manifold and valve seats (which were slightly increased in diameter). It was optionally tested without a catalytic converter. The technical specifications of the engine are summarized in Table 2.

Figure 2. TU5 JP4 engine and Telma CC-125 dynamometer.

Table 2. Technical specifications of the 1.6 L, 16v gasoline engine [21].

Engine designation	PSA TU5 JP4
Engine capacity (cm^3)	1587
Bore x Stroke (mm)	78.5 × 82.0
Number of valves	16
Compression ratio	10.8:1
Injection system	Multipoint
Maximum power (kW)	87 at 6600 rpm
Maximum torque (N.m)	145 at 5200 rpm

The test bench had an eddy current dynamometer (Telma CC-125) with electronic speed control to which this engine was attached (Figure 2). The engine was controlled by a 100% programmable electronic control unit from the ECU MASTER that enabled the injection and ignition to be mapped. A scale with resolution of 0.1 g (KERN FKB) was used to measure the fuel mass flow.

With the aid of an oxygen sensor (lambda probe—λ) adjustments were performed to the injection map until a stoichiometric air/fuel ratio (λ = 1) was reached for each fuel at all points on the map. Adjustments were also made to the ignition map (Figure 3), with the ignition being advanced until reaching the maximum effective torque point MBT (maximum braque torque) without knock.

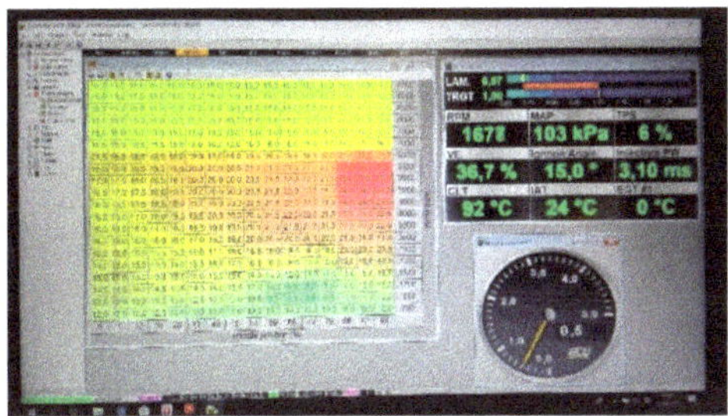

Figure 3. Ignition map.

The six fuels were tested at different speeds between 2000 and 6000 rpm in increments of 500 rpm. Most of the tests were performed at full load (100% throttle or wide-open throttle (WOT)), although some engine tests were carried out at partial loads (~50%) due to the amount of available pyrogasoline. Engine performance data and pollutants in the exhaust gases were recorded. The registrations occurred after stabilizing the engine (~20 s), and each registration lasted for 30 s.

To characterize the exhaust gas emissions, a gas analyser model AVL DIGAS 4000 Light was used. This equipment has five electrochemical sensors for measuring the compounds CO, CO_2, O_2, UHC, and NO_x, and the value of λ is calculated. The condensates were drained upstream of the analyser.

The engine performance analysis makes use of some measured or related parameters. The equations are shown in Supplementary Materials.

3. Results and Discussion

The first tests were performed with gasoline with 5% bio-oil (without distillation). These tests did not go well; they gave rise to a liquid purge through the head gasket and an intense odour in the laboratory. Additionally, there was a significant decrease in engine torque. Immediate discontinuation of the test prevented major engine problems. The disassembly and cleaning of the engine made it possible to identify the problem: The engine was not able to burn the heavier compounds of the bio-oil. This preliminary test definitively eliminated the use of non-distilled bio-oils in SI engines. All subsequent tests were only carried out with the light distilled part of the pyrolysis bio-oil (pyrogasoline). Kumar et al. [19] made the same conclusions, although for different reasons.

3.1. Fuel Characteristics

Table 3 and Figure 4 show the measured properties and the elementary compositions of the base fuels used.

Table 3. Density (ρ), higher heating value (HHV), lower heating value (LHV), pH, and stoichiometric air–fuel ratio (AFR) of the fuels used in the formulation of the mixtures.

Fuel	ρ (g.cm^{-3})	HHV (MJ.kg^{-1})	LHV (MJ.kg^{-1})	pH	AFR
Gasoline	0.75	42.6	40.0	-	14.2
Pyrogasoline	0.85	42.2	39.4	4.5	13.0
Ethanol	0.80	27.5	24.6	-	8.3

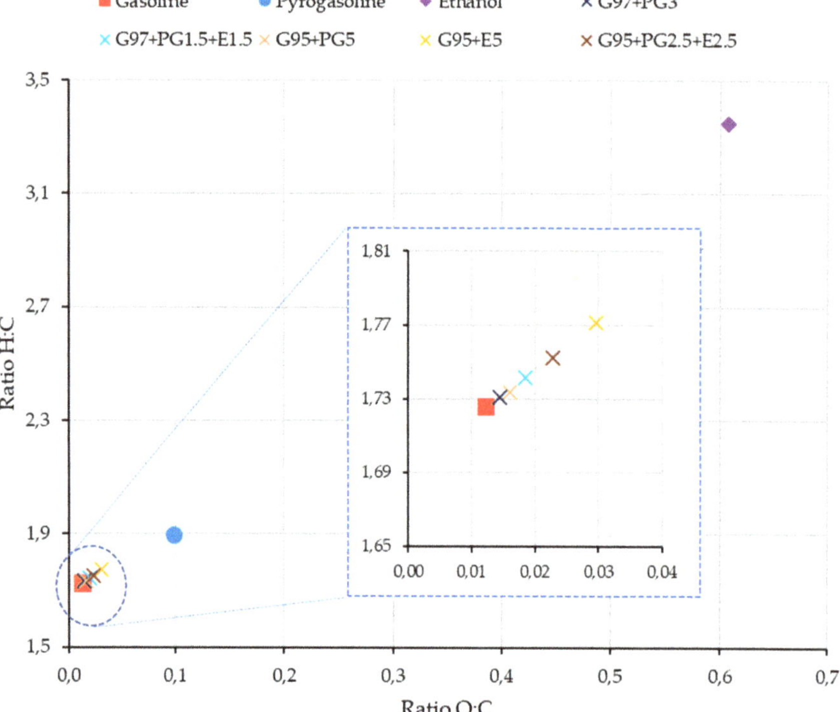

Figure 4. Elementary composition of the base fuels (gasoline, G; pyrogasoline, PG; and ethanol, E) and of the fuel mixtures.

Ethanol has a low energy content, less than two thirds of gasoline and pyrogasoline.

Binary and ternary mixtures of gasoline with pyrogasoline and ethanol in small percentages have practically the same O:C and H:C ratios as gasoline (Figure 4).

The functional groups of gasoline and pyrogasoline were identified by FT-IR. Table 4 shows the wave numbers of the main bands and functional groups assigned.

Table 4. Wave numbers (cm^{-1}) of the main bands of gasoline and pyrogasoline and the assigned functional groups [22,23].

Gasoline	Pyrogasoline	Functional Group/Assignment
3015	-	Alkene C-H strech
2957	2955	Methyl C-H asym./sym. stretch
2923	2922	Methylene C-H asym./sym. stretch
2870	-	Methyl C-H asym./sym. stretch
-	2853	Methylene C–H asym./sym. stretch
-	1710	Carboxylic acid C=O strech
1608, 1506 e 1490	-	Aromatic ring stretch
1456	1457	Methylene/Methyl C-H asym./sym. bend
1377	1377	Methyl C-H asym./sym. bend
-	1285	Aromatic ester C-O strech
1022	-	Aromatic C-H in-plane bend
-	909	Alkene C=C blend
878–698	722, 698	Aromatic C-H out-of-plane bend

Gasoline consists of light alkanes, few cycloalkanes, and many aromatic compounds. Pyrogasoline consists of alkanes, cycloalkanes, carboxylic acids, alkenes, and a few aromatics.

Pyrogasoline has less aromatics and branched alkanes than does gasoline; this difference in composition, according to Shamsul et al. [24], suggests that pyrogasoline has a lower octane number than gasoline.

Figure 5 presents the calculation of LHV and the air–fuel (AFR) for the fuels used in the tests.

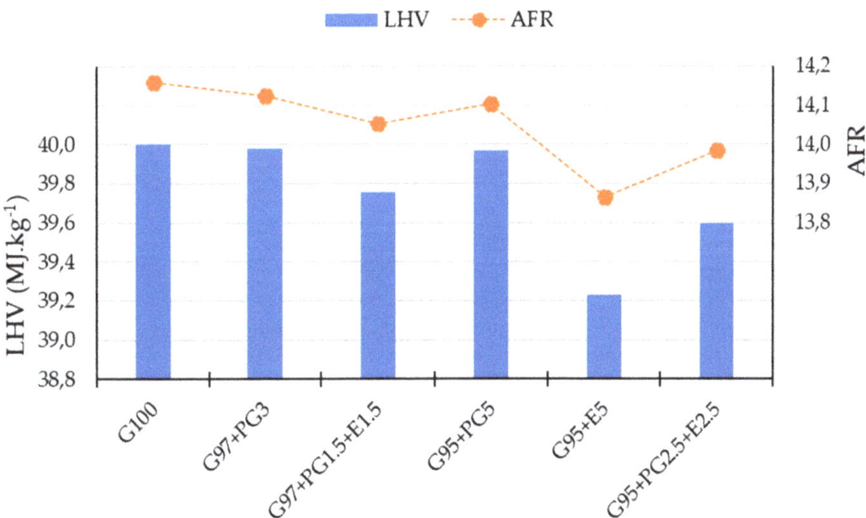

Figure 5. Lower heating value (LHV) and air–fuel ratio (AFR) for the fuels and fuel mixtures used in the engine tests.

3.2. Performance Analysis

The engine tests were performed on two consecutive days without major temperature or atmospheric pressure changes.

Figure 6 shows the torque and ignition advance curves of the six fuels used in the engine tests as a function of engine speed.

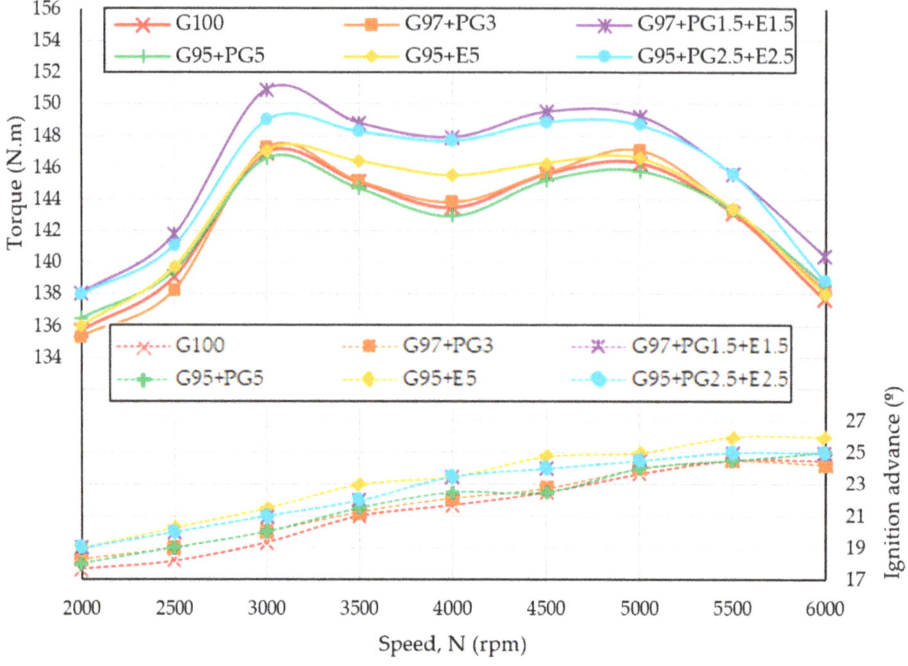

Figure 6. Torque (solid) and spark advance (dashed) curves for the different fuels.

The existence of different ignition advance curves shows that the different fuels properties influence the engine performance. The binary and ternary mixtures with ethanol have higher ignition advances than the rest of the fuels.

It appears that all torque curves showed two peaks, one close to 3000 rpm and the other close to 5000 rpm, as is normal in SI engines.

It was also verified that the six analysed fuels reached a maximum torque higher than that announced by the manufacturer. Gasoline reached a maximum torque of 146.3 N.m at 5000 rpm (instead of the 145 N.m at 5200 rpm announced), probably due to the referred small improvements introduced in the engine and/or due to the nonexistence of the catalytic converter.

One torque curve that is frankly the best: G97 + PG1.5 + E1.5. The other ternary mixture (G95 + PG2.5 + E2.5) is the second best, showing that the ternary mixtures positively improve the engine's performance. In the middle position is the binary mixture with ethanol G95 + E5. In the lower positions are the pump gasoline G100, the G97 + PG3, and the G95 + PG5, which are practically coincident. Binary mixtures with pyrogasoline alone did not seem to improve nor worsen the engine's performance compared to gasoline. The relative positioning of the G95 + E5 and G100 curves is in accordance with the literature [25]. Although the addition of ethanol slightly improved the base gasoline torque, it did not improve it as much as did the addition of pyrogasoline + ethanol.

Fuel G97 + PG1.5 + E1.5 had a slightly higher torque than that of G95 + PG2.5 + E2.5. The explanation may be found in the concentration of heavy pyrogasoline compounds that are solvated by the volatile components of gasoline.

Note that the blends containing ethanol are those with the best torque, which can be explained by the higher octane number of ethanol. They also allow the higher ignition advance without knock, which also confirms higher octane numbers.

Using the gasoline results as a comparison, Figure 7 shows the torque variations of the various fuels.

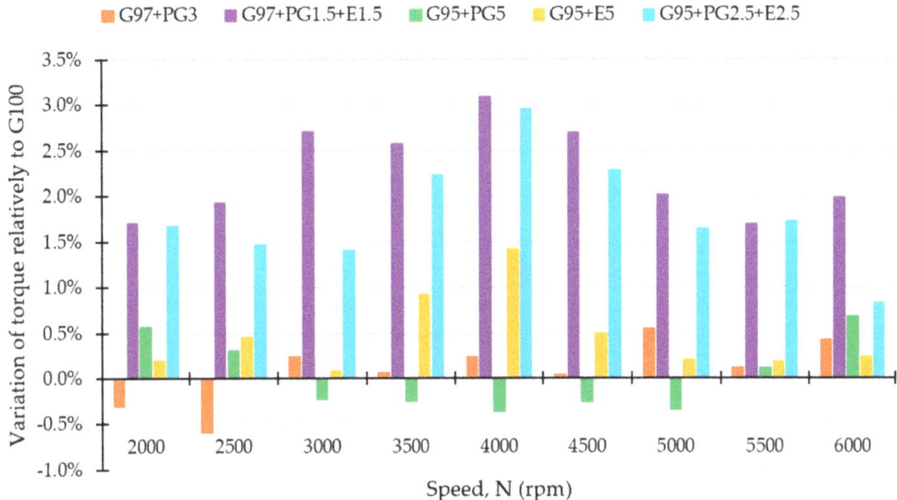

Figure 7. Torque variations (%) of the various fuel mixtures containing 97% and 95% gasoline in relation to the torque of 100% gasoline.

The results prove that the addition of small percentages of pyrogasoline and/or ethanol to commercial RON95 gasoline is beneficial for torque and power.

The ternary mixtures G97 + PG1.5 + E1.5 and G95 + PG2.5 + E2.5 present torque gains, between 0.8 and 3.1%, in relation to gasoline.

At low and high engine speeds, the torque difference between fuels is reduced.

After analysing the binary mixtures, we can conclude that:

(a) the engine worked well with all tested fuels;
(b) the incorporation of pyrogasoline did not negatively alter the engine torque;
(c) there is a synergistic gain in the presence of pyrogasoline and ethanol in small percentages in blends with gasoline.

Figure 8 shows the energy consumption for the different fuels along the engine speed range. As expected, the addition of ethanol reduced the energy content of the mixture, due to ethanol's lower LHV. The addition of pyrogasoline had the opposite effect, with the mixtures using it increasing their energy, as the lower stoichiometric air/fuel ratio of the pyrogasoline (Table 3) in relation to gasoline more than compensated for its lower LHV.

Figure 9 shows the brake specific energy consumption (BSEC) curves of the various fuels.

Through these specific consumption curves, two consumption minimums were identified at 3100 and 5000 rpm.

The binary mixture with 3% pyrogasoline had a consumption profile identical to that of 100% gasoline at most engine speeds, but when pyrogasoline incorporation was increased to 5%, an increase of BSEC occurred, especially in the range from 3000 to 5500 rpm, indicating that this biofuel was not being optimally burned. The binary mixture of 95% gasoline and 5% ethanol had the lowest brake specific energy consumption over the entire range of engine speeds, but the ternary mixture of 95% gasoline with 2.5% ethanol and 2.5% pyrogasoline also showed a consumption profile significantly lower than that of 100% gasoline.

Figure 10 shows the effective efficiency curves (η_e) of the various fuels.

The efficiency of the engine, using the different fuels, lies between 36 and 39% and, obviously, has opposite trends to the BSEC curves.

The efficiency curves closely follow the torque curves. As expected, the best average efficiency was achieved by G95 + E5 and the worst by G95 + PG5.

Using the results of gasoline as a baseline, Figure 11 shows the differences in the effective efficiency (η_e) of the various mixtures in relation to gasoline.

An analysis of Figure 11 shows that the efficiency differences are small, but the fuel blends with 2.5% and 5% ethanol showed improvements of 0.4% to 1.2%, respectively, of engine efficiency over the speed range. Although revealing the highest engine efficiency, the mixtures containing ethanol did not display the highest torque (Figure 6). The reason is the lower energy content of the mixture, as shown in Figure 8.

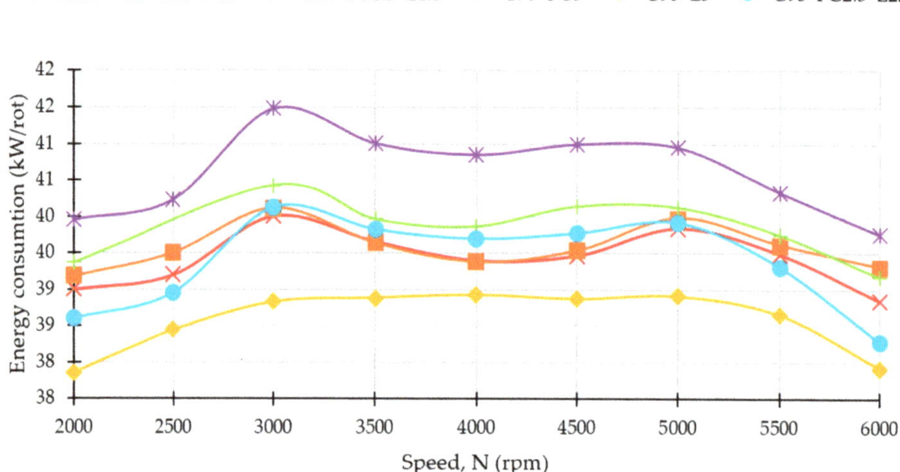

Figure 8. Energy consumption for the different fuels.

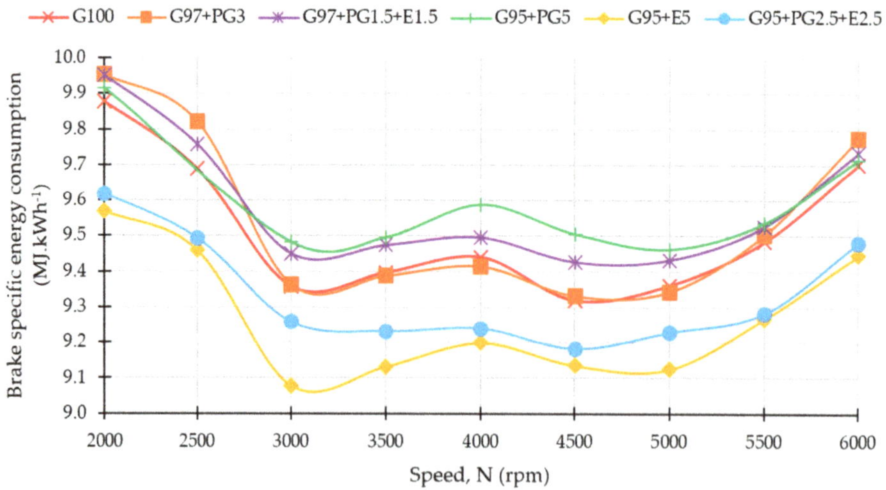

Figure 9. Brake specific energy consumption (BSEC) for the different fuels.

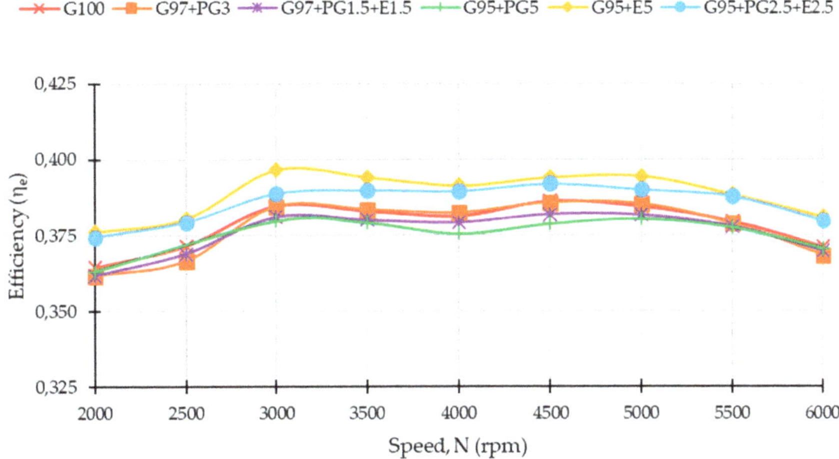

Figure 10. Effective efficiency curves of the fuels used in the tests.

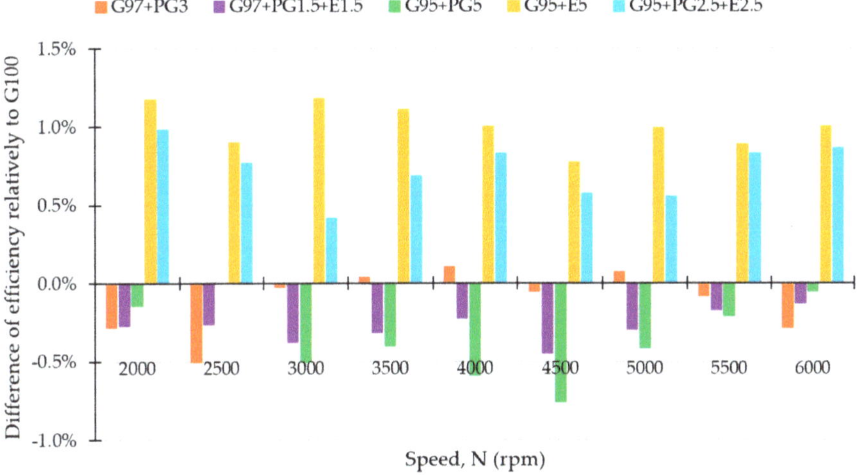

Figure 11. Differences in the efficiency of fuel mixtures in relation to gasoline.

3.3. Emission Analysis

Although values for all rotation speeds were recorded, all points where the condition $\lambda = 1.00$ was not observed were removed from the graph. Figure 12 shows the CO emissions from the fuels used in the tests.

CO emissions follow a W pattern as can be seen in Figure 12. The CO graph has two minimums that coincide with the maximum torque, which makes sense; these two points are also the ones with the highest engine efficiency. The engine burns well at these speeds.

The G95 + E5 and G97 + PG1.5 + E1.5 fuels achieved the lowest and highest CO emissions, respectively. This difference in behaviour may be due to the higher viscosity of pyrogasoline, which worsens fuel atomization and leads to incomplete combustion.

Figure 13 shows UHC emissions of the fuels used in the tests.

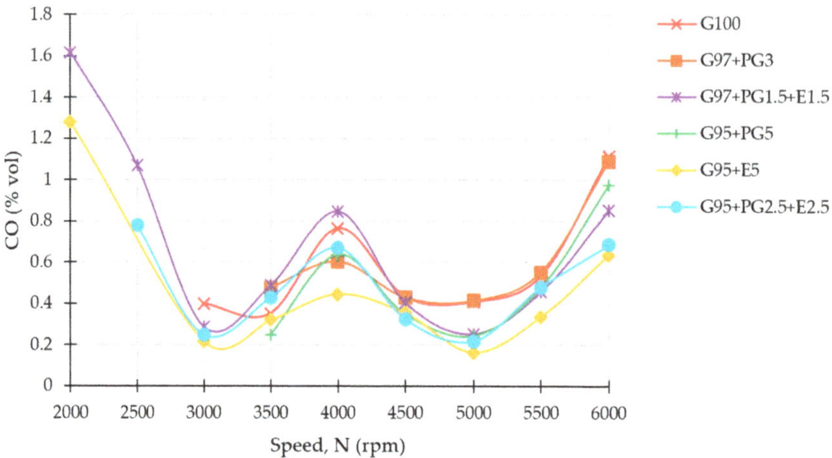

Figure 12. CO emissions of fuels used in the tests.

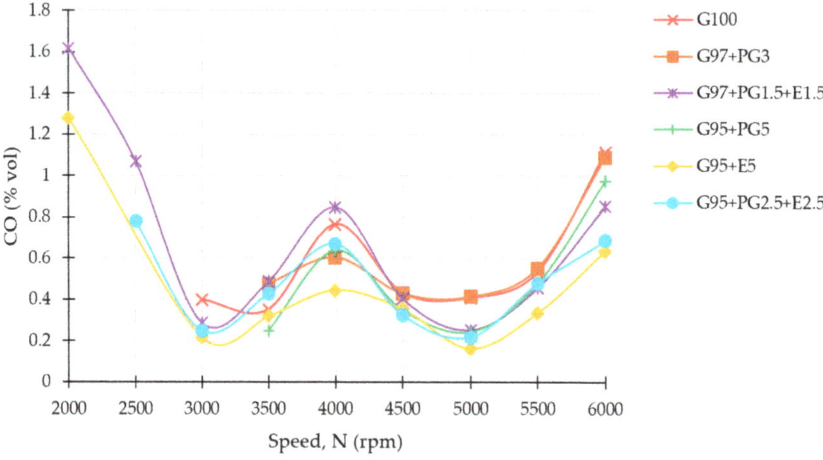

Figure 13. Unburned hydrocarbon (UHC) emissions of the fuels used in the tests.

Again, points that did not meet the criterion $\lambda = 1.00$ are not shown in the graph. The various points do not follow specific trends of the curves for torque and for CO emissions, so trend curves approximated by 2nd degree polynomials were used, which seems to make more sense.

The design of the UHC curves seems to show higher values for low and for high engine speed, while displaying lower values for intermediate speeds. In order to try to reveal more meaningful trends, power was used instead of engine rpm, as it may be more related to emission production, although the results were similar. So, the weak turbulence (lower speeds) and reduced combustion time (higher speeds) seems to be responsible for increasing the UHC emissions in the exhaust gases.

Gasoline displays the lowest UHC curve while binary mixtures with pyrogasoline (G97 + PG3 and G95 + PG5) have the highest UHC curves, so it seems that the addition of pyrogasoline increases the emission of that pollutant.

Figure 14 shows NO_x emissions from the various fuels used in the tests. The conditions used for Figure 13 were also used in Figure 14.

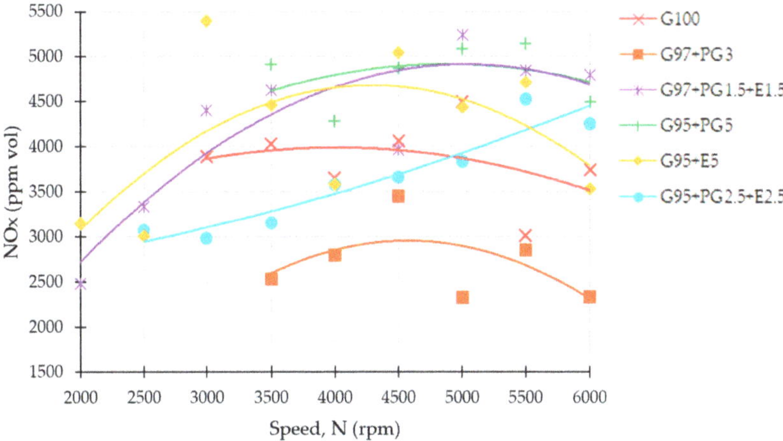

Figure 14. NO_x emissions for the fuels used in the tests.

In general, NO_x emissions seem to increase with engine speed, as the power also increases, and NO_x emissions are related to gas temperature.

The very high NO_x values are probably related with the changes made to the engine that improved its performance, but additionally increased NO_x.

The longer the ignition advance, the higher is the pressure and temperature achieved by the combustion products that increase NO_x production [25]. In general, the ignition advance is limited by the production of NO_x, but this trend was not seen here. The advance of the ignition was only limited by the knock onset and the torque maximization. Therefore, minimal changes of the ignition advance may show larger excursions of NO_x emissions, which may explain the lack of specific trends for this pollutant.

It was supposed that oxygenates present NO_x emissions exceeding those of the non-oxygenates [14], but this trend has not been fully verified, as NO_x emissions above and below those of gasoline have been verified with the addition of ethanol. Again, small changes of the ignition advance may result in larger changes than those predicted by the use of oxygenates.

It was not possible to notice a specific behaviour of fuels with pyrogasoline added in relation to NO_x: G95 + PG5 showed an increase in relation to RON95 gasoline (being practically the mixture that presented the highest NO_x), but G97 + PG3 was the mixture that produced lower levels of NO_x. Overall, we can state that the NO_x emissions do not show an obvious trend in terms of pyrogasoline use as an additive to gasoline.

4. Conclusions

Although straight pyrolysis bio-oil should not be used in spark ignition engines (due to the presence of soluble solids and high molecular weight liquids), the light distillates (pyrogasoline) of this bio-oil did prove adequate for use in SI engines.

The characterization of the light fraction of distillate pyrolysis bio-oil points to a similar composition to that of gasoline and comparable LHVs.

Binary mixtures with pyrogasoline or ethanol did not improve or worsen the engine's performance, but ternary mixtures with small percentages of pyrogasoline and ethanol positively improved the engine's performance, with torque gains between 0.8 and 3.1% compared to commercial gasoline RON95.

Ethanol mixtures showed the highest efficiency values, whereas mixtures with pyrogasoline had the lower values. The best thermal efficiency was achieved by the binary mixture of 95% gasoline and 5% ethanol and the worst by the binary mixture of 95% gasoline and 5% pyrogasoline.

Mixtures with oxygenated fuels showed reductions in CO values. NO_x emissions from oxygenated fuels had contradictory behaviours in relation to gasoline. The authors could not verify a trend (of increase or decrease) in the pyrogasoline addition to gasoline in terms of NO_x production.

The results achieved by pyrogasoline seem to accompany those of other additives and biofuels currently used in engines. The tests showed that the use of pyrogasoline is viable as an alternative or as a complement to ethanol, contributing to the increase of the renewable fraction of the fuels in spark ignition engines.

The legislative limitations to supplementing gasoline with first generation ethanol may favour the use of combined additives (ternary mixtures) as used in this work. Furthermore, the use of ternary mixtures of gasoline with oxygenated and non-oxygenated biofuels may allow the achievement of a compromise between engine performance and combustion emissions while contributing to a diversification of the biofuel component. The results obtained in this work also validate an alternative use for low quality lipids as feedstocks to the production of advanced biofuels.

Supplementary Materials: The following are available online at http://www.mdpi.com/1996-1073/13/18/4671/s1.

Author Contributions: Conceptualization, M.G.; Data curation, L.D., F.P.B., and J.M.; Formal analysis, L.D. and J.M.; Investigation, L.D., J.C., and T.A.; Methodology, J.C., T.A., and J.M.; Project administration, M.G.; Resources, M.G.; Supervision, F.P.B. and M.G. All authors have read and agreed to the published version of the manuscript.

Funding: This work was partially funded by Fundação para a Ciência e Tecnologia: UIDB/04077/2020.

Conflicts of Interest: The authors declare no conflict of interest.

Glossary

ρ	Density
AFR	Air–fuel ratio
BSEC	Brake specific energy consumption
CO	Carbon monoxide
CO_2	Carbon dioxide
CNG/LPG	Compressed natural gas/Liquid petroleum gases
DPPO	Distilled PPO
E	Ethanol
GHG	Greenhouse gas
G	Gasoline
GLF	Gasoline-like fuel
HC	Hydrocarbons
HHV	Higher heating value
H:C	Hydrogen-to-carbon ratio
LHV	Lower heating value
m	Mass
MBT	Maximum braque torque
NO_x	Oxides of nitrogen
O:C	Oxygen-to-carbon ratio
ON	Octane number
PG	Pyrogasoline
PPO	Pyrolysis bio-oils of non-distilled plastic waste
SI	Spark ignition
UHC	Unburned hydrocarbons
WOT	Wide open throttle

References

1. Cazzola, P. In Proceedings of the Insights Emerging from the 2015 Global EV Outlook (IEA), Goyang, Korea, 4 May 2015. Available online: https://www.iea.org/events/towards-a-global-ev-market (accessed on 4 September 2020).
2. Coelho, P.; Costa, M. *Combustão*; Edições Orion: London, UK, 2007; p. 714.
3. IEA. *Breakdown of Sectoral Final Consumption by Source*; IEA: Paris, France, 2011.

4. Annex, A. *Organisation for Economic Co-Operation. Agrcultural Outlook 2008–2017*; OECD-FAO: Paris, France, 2008; pp. 1–73.
5. IEA. Technology Roadmap-Biofuels for Transport. 2011. Available online: papers2://publication/uuid/7E683CA3-E72A-439B-8C06-CDAADB021565 (accessed on 4 September 2020).
6. EC. Renewable Energy-Recast to 2030 (RED II). Available online: https://ec.europa.eu/jrc/en/jec/renewable-energy-recast-2030-red-ii (accessed on 5 August 2020).
7. Directive 2018/2001/EU. Directive (EU) 2018/2001 of the European Parliament and of the Council on the Promotion of the Use of Energy from Renewable Sources. *Off. J. Eur. Union* **2018**, *2018*, 82–209. Available online: https://eur-lex.europa.eu/legal-content/EN/TXT/PDF/?uri=CELEX:32018L2001&from=EN (accessed on 4 September 2020).
8. Comission of the European Communities. *GREEN PAPER On the Management of Bio-Waste in the European Union*; Comission of the European Communities: Brussels, Belgium, 2008; p. 19.
9. European Commission. Directive 2015/1513 of the European parliament and of the council. *J. Eur. Union.* **2015**, *2014*, 20–30.
10. Asomaning, J.; Mussone, P.; Bressler, D.C. Thermal deoxygenation and pyrolysis of oleic acid. *J. Anal. Appl. Pyrolysis* **2014**, *105*, 1–7. [CrossRef]
11. Billaud, F.; Minh, A.K.T.; Lozano, P.; Pioch, D. Étude paramétrique du craquage catalytique de l'oléate de méthyle. *Comptes Rendus Chim.* **2004**, *7*, 91–96. [CrossRef]
12. Demirbas, A. Gasoline-rich Liquid from sunflower oil by catalytic pyrolysis with alumina-treated sodium hydroxide. *Energy Sources Part A Recover. Util. Environ. Eff.* **2009**, *31*, 671–678. [CrossRef]
13. Roesyadi, A.; Hariprajitno, D.; Nurjannah, N.; Savitri, S.D. HZSM-5 catalyst for cracking palm oil to gasoline: A comparative study with and without impregnation. *Bull. Chem. React. Eng. Catal.* **2013**, *7*, 185–190. [CrossRef]
14. Agarwal, A.K. Biofuels (alcohols and biodiesel) applications as fuels for internal combustion engines. *Prog. Energy Combust. Sci.* **2007**, *33*, 233–271. [CrossRef]
15. Zhao, F.; Lai, M.-C.; Harrington, D.L. Automotive spark-ignited direct-injection gasoline engines. *Prog. Energy Combust. Sci.* **1999**, *25*, 437–562. [CrossRef]
16. Bušić, A.; Marđetko, N.; Kundas, S.; Morzak, G.; Belskaya, H.; Šantek, M.I.; Komes, D.; Novak, S.; Šantek, B. Bioethanol production from renewable raw materials and its separation and purification: A review. *Food Technol. Biotechnol.* **2018**, *56*, 289–311. [CrossRef] [PubMed]
17. Öztop, H.F.; Varol, Y.; Altun, Ş.; Firat, M. Using gasoline-like fuel obtained from waste automobile tires in a spark-ignited engine. *Energy Sources Part A Recover. Util. Environ. Eff.* **2014**, *36*, 1468–1475. [CrossRef]
18. Suiuay, C.; Laloon, K.; Katekaew, S.; Senawong, K.; Noisuwan, P.; Sudajan, S. Effect of gasoline-like fuel obtained from hard-resin of Yang (Dipterocarpus alatus) on single cylinder gasoline engine performance and exhaust emissions. *Renew. Energy.* **2020**, *153*, 634–645. [CrossRef]
19. Kareddula, K.; Puli, R. Influence of plastic oil with ethanol gasoline blending on multi cylinder spark ignition engine. *Alex. Eng. J.* **2018**, *57*, 2585–2589. [CrossRef]
20. Kumar, K.V.; Puli, R.K.; Kumari, A.S.; Shailesh, P. Performance and emission studies of a SI engine using distilled plastic pyrolysis oil-petrol blends. *MATEC Web Conf.* **2016**, *45*, 3002. [CrossRef]
21. Owners Club 206cc P. TU5JP4 Technical Specifications. Peugeot 206cc Owners Club. Available online: http://www.peugeot206cc.co.uk/repair-206/206/info/gb/b1bbmek3.htm (accessed on 1 April 2019).
22. Coates, J. Interpretation of infrared spectra, a practical approach. *Encycl. Anal. Chem.* **2006**, 10815–10837. [CrossRef]
23. OChemOnline. Infrared spectroscopy absorption table. *Chem. Libr.* **2016**, 1–4. Available online: https://chem.libretexts.org/Bookshelves/Ancillary_Materials/Reference/Reference_Tables/Spectroscopic_Parameters/Infrared_Spectroscopy_Absorption_Table (accessed on 4 September 2020).
24. Shamsul, N.S.; Kamarudin, S.K.; Rahman, N.A. Conversion of bio-oil to bio gasoline via pyrolysis and hydrothermal: A review. *Renew. Sustain. Energy Rev.* **2017**, *80*, 538–549. [CrossRef]
25. Martins, J. *Motores de Combustão Interna*, 6th ed.; Engebook: Porto, Portugal, 2020; p. 586. ISBN 9789898927842.

Article

Effects of Diethyl Ether Introduction in Emissions and Performance of a Diesel Engine Fueled with Biodiesel-Ethanol Blends

Márcio Carvalho [1,2,*], Felipe Torres [1,3], Vitor Ferreira [4], Júlio Silva [5], Jorge Martins [6] and Ednildo Torres [7]

1 Polytechnic School, Industrial Engineering Postgraduate Program, Federal University of Bahia, Rua Aristides Novis, 2, Federação, Salvador 40210-630, Brazil; ftorres@ufrb.edu.br
2 Multidisciplinary Campus of Bom Jesus da Lapa, Federal University of West of Bahia, Av. Manoel Novaes, 1064, Bom Jesus da Lapa 47600-000, Brazil
3 Department of Mechanical Systems, Federal University of Reconcavo of Bahia, Rua Rui Barbosa, 710, Centro, Cruz das Almas 44380-000, Brazil
4 Institute of Science, Technology and Innovation, Federal University of Bahia, Rua do Telegráfo S/N, Camaçari 42809-000, Brazil; vitorpf@ufba.br
5 Department of Mechanical Engineering, Federal University of Bahia, Rua Aristides Novis, 2, Federação, Salvador 40210-630, Brazil; julio.silva@ufba.br
6 Department of Mechanical Engineering, Minho University, Av. da Universidade, Guimarães 4800-058, Portugal; jmartins@dem.uminho.pt
7 Department of Chemical Engineering, Federal University of Bahia, Rua Aristides Novis, 2, Federação, Salvador 40210-630, Brazil; ednildo@ufba.br
* Correspondence: marcio.carvalho@ufob.edu.br

Received: 2 June 2020; Accepted: 29 June 2020; Published: 23 July 2020

Abstract: Biofuels provide high oxygen content for combustion and do modify properties that influence the engine operation process such as viscosity, enthalpy of vaporization, and cetane number. Some requirements of performance, fuel consumption, efficiency, and exhaust emission are necessary for the validation of these biofuels for application in engines. This work studies the effects of the use of diethyl ether (DEE) in biodiesel-ethanol blends in a DI mechanical diesel engine. The blends used in the tests were B80E20 (biodiesel 80%-ethanol 20%) and B76E19DEE5 (biodiesel 76%-ethanol 19%-DEE 5%). Fossil diesel (D100) and biodiesel (B100) were evaluated as reference fuels. The results revealed similar engine efficiencies among tested fuels at all loads. The use of B100 increased CO and NO_x and decreased THC compared to D100 at the three loads tested. B80E20 fuel showed an increase in NO_x emission in comparison with all fuels tested, which was attributed to higher oxygen content and lower cetane number. THC and CO were also increased for B80E20 compared to B100 and D100. The use of B76E19DEE5 fuel revealed reductions in NO_x and CO emissions, while THC emissions increased. The engine efficiency of B76E19DEE5 was also highlighted at intermediate and more elevated engine load conditions.

Keywords: Biodiesel; diesel engines; diethyl ether; ethanol; biofuels; emissions

---

1. Introduction

The growing concern over climate change and fossil fuel dependency has increased visibility for renewable energy sources [1]. Biofuels have been identified as promising renewable fuels, in particular, when originated from waste feedstock and non-edible plant species [2].

The use of alternative fuels in compression ignition (CI) engines should be evaluated in many terms, as exhaust emissions, fuel stability, availability, distribution, and impacts on engine durability [3].

Among biofuels, biodiesel is considered a promisor fuel substitute due to similar results of engine performance and efficiency with those obtained with neat diesel. However, the elevated viscosity, higher cloud point, and pour point properties are adverse factors that can result in solidification during cold weather, causing clogging in filter and fuel lines that lead to engine damage [4]. The addition of less viscous fuels such as ethanol or diethyl ether (DEE) into biodiesel or diesel-biodiesel blends can improve the fuel spray characteristics in the combustion chamber and avoid clogging problems [5–7]. One of the main objectives of the research work with alternative fuels focused on optimizing the fuel blend based on fuel properties like kinematic viscosity, density, cloud point, and pour point [8]. Besides that, studies considering the feasibility of blending biofuels with fossil fuels in terms of heating and evaporation are important. In this context, Al-Esawi, Qubeissi, and Kolodnytska [9] reported that pure biodiesel and pure ethanol had 11.7% and 43.3% less droplet lifetime than pure diesel, ascribed to the fact that ethanol and biodiesel had higher vapor pressures than diesel. The droplet lifetime also decreased in relation to diesel when fractions of biodiesel, ethanol, or both fuels were used in blends with diesel. However, the differences were less than 2%. A similar lifetime study was carried out by Al Qubeissi et al. [10] using the gasoline fossil in comparison to ethanol. However, the results showed that, in this case, gasoline had an average time of about 34% less than pure ethanol. A complementary study by Al-Esawi et al. [11] used E85 (85% ethanol and 15% fossil gasoline) in blends with diesel. Results showed that the droplet lifetime for pure diesel was longer than that for any blend. The difference reaches 49.5% for pure E85 and was about 6% for the blend E85 with 5% of diesel.

The use of ethanol in blends with diesel is also justified by the percentage increase of biofuel in the blend and by technical issues, such as increased oxygen content and the possibility of reducing pollutant emissions, such as particulate matter and NO_x, simultaneously [6]. Some properties of ethanol, however, are adverse in terms of CI engine requirements. For example, its low cetane number (CN) makes it unfeasible to be used as the main fuel in CI engines. On the other hand, ethanol can be used blended with diesel or biodiesel and can result in the improvement of volumetric efficiency and reduction of particulate matter (PM) emission. In general, the low CN of ethanol induces a longer ignition delay, resulting in more premixed mixture and higher heat release rate (HRR) during combustion process [7]. However, the higher enthalpy of vaporization of ethanol produces a cooling effect that results in reduction of the global combustion temperature and consequently NO_x formation decrease. In this way, opposite results regarding NO_x emission due to ethanol addition can be found in literature. Results may vary according to blend composition, engine design, and method of application. Tutak et al. [12] developed a comparative study of the effect of the use of diesel-ethanol and biodiesel-ethanol blends on performance and emission characteristics of a diesel engine. The tests were conducted at a constant angle of diesel fuel injection, full load, and constant rotational speed (1500 rpm). Authors observed thermal efficiency increase using high ethanol content in diesel-biodiesel blends, while there was similar thermal efficiency in biodiesel-ethanol blends. Authors observed thermal efficiency increase using high ethanol content in diesel-biodiesel blends, while similar thermal efficiency in biodiesel-ethanol blends was observed. Emissions of total hydrocarbons (THC) and NO_x, however, were higher for the blends when compared with neat diesel and biodiesel. Yilmaz [13] observed a reduction in NO_x emissions when using a blend of biodiesel (85%) and ethanol (15%) in comparison to diesel and biodiesel. The author attributed this result to the higher enthalpy of vaporization and the lower heating value (LHV) of the ethanol, which reduced the combustion temperature. Engine tests were developed at part-load and full-load conditions at a constant speed. In a study by Prbakaran and Viswanathan [14], blends of cottonseed oil methyl ester and anhydrous ethanol in 10%, 30%, and 50% in volume were tested in a diesel engine at various loads. It was observed that NO_x emission was reduced in blend containing 50% ethanol, for all the loads, comparing to the fossil diesel and the other blends. The evaporation characteristic of B50E50 was reported as one of the reasons for the reduction of NO_x. The use of the blends B70E30 and B90E10, however, showed an increase in NO_x at higher loads, which was attributed to combustion temperature increase by the additional oxygen content promoted by ethanol addition. Kandasamy et al. [15] observed that the use of 20% of ethanol in a B5 blend (5% of esterified cotton seed methyl ester and 95% of neat diesel) decreased NO_x and unburned hydrocarbon emissions at lower to medium speed range and increased thereafter.

The research work with alternative fuels focused on optimizing the fuel blend based on fuel properties like kinematic viscosity, density, cloud point, and pour point.

DEE appears as a potential additive to diesel-biodiesel-ethanol blends among oxygenated biofuels due to its chemical properties such as high cetane number (>125), moderate energy density (similar to biodiesel), high oxygen content in its structure, low self-ignition temperature, prolonged flame duration, and adequate miscibility with diesel, biodiesel, and ethanol. In this context, DEE can be considered more suitable to be used for CI engines application than ethanol or methanol due to its higher CN and LHV [16]. Some research has been developed using DEE to improve ignition quality in blends composed by diesel and other feedstocks fuels. Lee and Kim [17] evaluated blends of diesel and DEE and results showed similar engine efficiency compared to diesel along with lower emissions of THC, CO, and PM. However, NO_x emissions were higher. The authors attributed the results to the shorter ignition delay and high oxygen content of the blends. Qi et al. [18] investigated the effects of ethanol and DEE as additives in a diesel-biodiesel blend. The tested fuels were B30 (30% biodiesel and 70% diesel), BE-1 (5% diethyl ether, 25% biodiesel, and 70% diesel) and BE-2 (5% ethanol, 25% biodiesel, and 70% diesel). Reduction in smoke and CO were observed with BE-1 and BE-2, while NO_x was higher for BE-2 and HC was higher for BE-1 and BE-2 when compared to B30. Jeevanantham et al. [19] used DEE in blends with diesel-biodiesel in volumetric proportion of 5% (D50B45DEE5) and 10% (D50B40DEE10). Results showed that NO_x emissions were reduced in all test conditions concerning the other fuels. The authors attributed the results to the significant cooling effect caused by the high enthalpy of vaporization of DEE. This work approaches the application of DEE in biodiesel-ethanol blends. The DEE was used toward enhancing the CN, which was deteriorated as a result of ethanol introduction. In the literature, it is common to use blends with DEE, however, most part considering diesel in its composition. This work evaluates the performance and emissions of a light-duty naturally aspirated diesel engine with mechanical fuel injection using biodiesel, ethanol, and DEE. DEE is used as an additive to increase the CN, which is deteriorated as a result of ethanol introduction. The tests were developed in a light-duty naturally aspirated diesel engine with mechanical fuel injection, evaluate the performance, efficiency, and exhaust emissions.

2. Materials and Methods

2.1. Experimental Setup

The engine used in the experiments was a four-stroke, two cylinders, direct fuel injected, and naturally aspirated diesel engine. The main characteristics of the engine are summarized in Table 1.

Table 1. Engine characteristics.

Engine Manufacturer	Agrale
Model	M790
Number of cylinders	2
Bore × Stroke	90 mm × 100 mm
Engine displacement	1272 cm^3
Compression ratio	18:1
Maximum Brake Power (kW)	19.8 kW at 3000 rpm
Maximum torque (Nm)	70 Nm at 2250 rpm
Injection type	Direct Injection.

The engine original specifications were kept constant throughout the tests.

Figure 1 shows a schematic diagram of the experimental setup that consists of a diesel engine coupled with a hydraulic dynamometer (Shenck D210), a gas analyzer for exhaust emission, HC meter, fuel tank, scale, high-pressure fuel pump, and K-type thermocouples for temperature measurement.

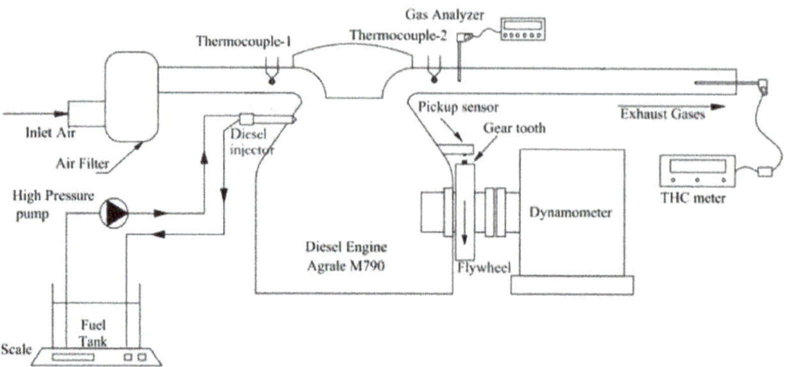

Figure 1. Schematic diagram of experimental setup.

Experimental tests were performed under three loads: 2.7 kW, 5.4 kW, and 8.1 kW, and engine speed was fixed at 1700 rpm. These loads correspond to 25%, 50%, and 75% of the maximum brake power (BP) at the engine speed of 1700 rpm, using mineral diesel as a reference fuel. These loads were chosen since engine maximum load is not achievable with blends of reduced LHV.

The amount of fuel injected varied for each fuel tested due to the difference in LHV between the fuels (Table 2). This control was performed by varying the acceleration in the high-pressure fuel injection pump.

Table 2. Compositions of tested fuels.

Fuel	Diesel	B100	Ethanol	DEE	LHV (MJ/kg)
D100	100%	0%	0%	0%	42.50
B100	0%	100%	0%	0%	37.45
B80E20	0%	80%	20%	0%	35.80
B76E19DEE5	0%	76%	19%	5%	35.84

2.2. Fuels Characteristics

Evaluated blends were prepared using soybean biodiesel (B100), ethanol (purity of 99.3%), and diethyl-ether (DEE, purity of 99.7%). Diesel fossil pure (D100, maximum 10 ppm of sulfur) and pure biodiesel were tested as reference. The main characteristics of the fuels are depicted in Table 3.

Table 3. Properties of original fuels.

Properties	Diesel	Biodiesel	Ethanol	DEE
Formula	$C_{10}H_{18}$	$C_{18}H_{34}O_2$	C_2H_6O	$C_4H_{10}O$
Oxygen content, (%)	0	10.8	34.7	21.6
Density @ 20 °C, (g/m^3)	0.840	0.878	0.786	0.713
Viscosity @ 40 °C, (cSt)	3.30	4.95	1.20	0.23
Flash point, (°C)	96	158	15	−45
Cetane number	46	56	6.5	120
Lower heating value, (MJ/kg)	42.50	37.45	28.40	36.87
Enthalpy of evaporation, (kJ/kg)	260	200	836	356

An IKA C2000 bomb calorimeter was employed to determine the lower heating value based on the standard ASTM-D240-87. The fuel density was measured using a DMA-5000 densimeter, while samples viscosity was measured by a model P capillarity viscometer. Table 2 shows the composition of single fuels, biodiesel-ethanol, and biodiesel-ethanol-DEE along with their volumetric compositions and LHV.

The LHV was predicted based on the volumetric fractions, density, and energy fractions of each blend component. This methodology was also used by Al-Esawi, Al Qubeissi, and Kolodnytska [9].

The stability of the B76E19DEE5 blend was evaluated under the test temperatures to certify that no fuel would be lost by evaporation.

2.3. Instrumentation

Fuel mass flow rate was obtained by gravimetric method using a digital scale while fuel consumption was evaluated under five cycles of measurement with a sampling time of 30 minutes. Exhaust gas emissions of CO, NO_X, and THC were assessed using two gas analyzers under an average of 30 measurements for each fuel. The main characteristics and uncertainty of the employed instruments are shown in Table 4.

Table 4. Properties of the main instruments.

Measure	Instrument	Manufacturer (model)	Range	Uncertainty
Ambient humidity	Digital hygrometer	Icel (HT-208)	0 to 100%	± 3%
Fuel consumption	Digital scale	Mettler Toledo (9094)	0 to 15 kg	± 2%
Exhaust Gas (NO_X; CO)	Gas analyzer	COSA (Optima 7)	0–1000 ppm	± 5%
Exhaust Gas (THC)	Gas analyzer	NAPRO (PC Multigas)	0–2000 ppm	± 5%

In order to enhance the confidence level of the experiments, the tests were carried out under similar system conditions for all fuels, such as engine temperature and weather temperature between (29 ± 2 °C) and humidity (55 ± 8%).

3. Results and Discussions

3.1. Engine Performance

3.1.1. Brake Specific Fuel Consumption (BSFC)

Brake specific fuel consumption may be defined as a ratio between fuel mass flow rate and engine brake power, which usually depends on the volumetric fuel injection system and fuel properties (e.g., density, calorific value, and viscosity). Figure 2 outline an increase in BSFC when the engine operates with biofuels compared to mineral diesel (D100), which may be explained by the lower LHV of biofuels blends, as shown in Table 2.

It is also observed that as loads increases, BSFC decreases for all fuels due to the higher engine efficiency at high engine loads [20].

When engine operated with B100 there was an increase in BSFC of about 14% on average, under the three loads, in comparison with D100, while the difference in LHV was only 12% lower for B100 in comparison with D100. Thus, it is possible to infer a slight reduction in engine efficiency when B100 is used. Besides, the higher viscosity of B100 leads to an increase in BSFC. Higher viscosity alters fuel jet atomization, increasing the average diameter and density of fuel droplets while reduces combustion efficiency [21]. When compared to B100, the biodiesel-ethanol (B80E20) blend has shown an increase in BSFC of 5.3%, 4.2%, and 3% at loads of 2.7 kW, 5.4 kW, and 8.1 kW, respectively. The further increase of fuel consumption of B80E20 is due to a reduction in LHV of about 4.4% for the blend, which occurred as a result of the introduction of ethanol, which explains the increase in BSFC. Furthermore, a greater difference can be observed at lower loads due to lower combustion temperatures [22], whereas the high ethanol enthalpy of vaporization induces a reduction in combustion temperature and thus combustion efficiency decreases. A slight decrease in BSFC was observed in the blend with diethyl ether (B76E19DEE5) in concerning to B80E20 blend, under all tested conditions. This reduction can be attributed to the short increment in LHV of about 1.2%. However, other factors also contribute to improving combustion quality for this blend, such as higher CN and the high oxygen content [8].

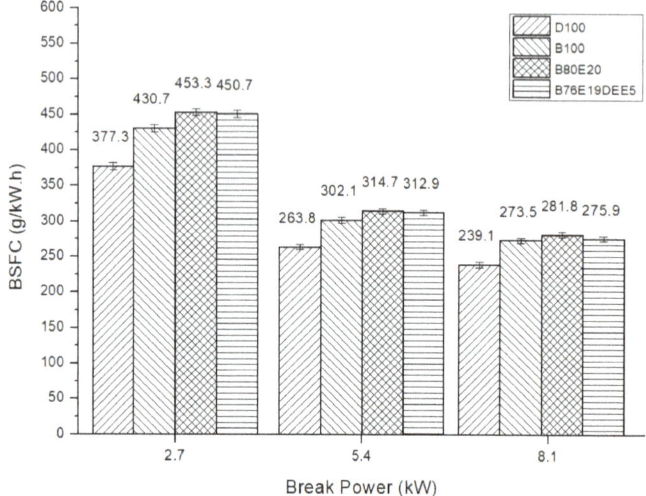

Figure 2. The variation of brake specific fuel consumption with respect to the engine load.

3.1.2. Brake Specific Energy Consumption (BSEC)

Another way of evaluating the performance of the fuels tested is by using the BSEC. While BSFC represents a measure of fuel mass consumption, BSEC allows a view of the energy consumption of the fuels, in accordance with the engine brake power. The results of the BSEC obtained in the experiments are presented in Figure 3.

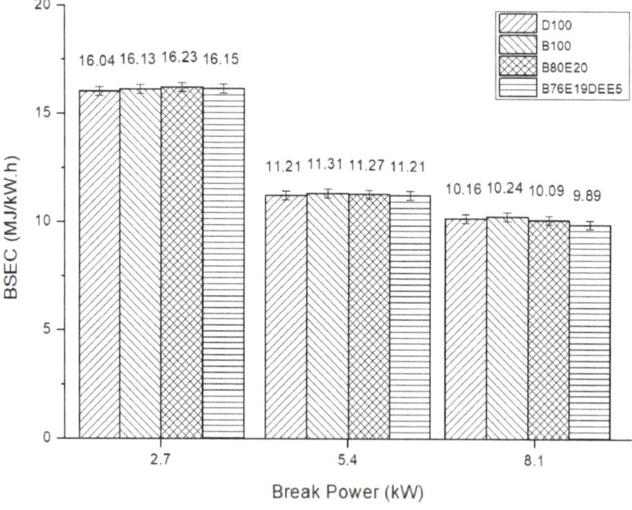

Figure 3. The variation of brake specific energy consumption with respect to the engine load.

Figure 3 shows relatively small differences between the fuel samples tested. This similarity of results shows that the performance of the proposed fuel blends is consistent with the conventional diesel of the engine design. Considering the test conditions, in the lowest load, 2.7 kW, the lowest BSEC value can be observed with the D100 and the highest value for the B80E20 fuel. This can be attributed to the cooling effect of the chamber and, therefore, to the increase of the delay of the start of the ignition using B80E20. These factors are due to the high enthalpy of evaporation and low CN of the blend B80E20, which has more influence at the low load condition due to the lower temperatures in the chamber wall and in the residual gases of the combustion [12,15]. Regarding the D100, its better results can be explained by adequate properties of the D100 in accordance with the design parameters of the diesel engine, providing adequate conditions for burning [20].

At the highest load, 8.1 kW, an improvement of the BSEC was observed using B80E20 with respect to D100. This can be explained due to the higher temperatures in the chamber and in the residual gases of combustion in that condition. In addition, the fuel oxygen content improves the fuel-burning quality [8,12,15].

Considering B76E19DEE5, the presence of DEE reduced the BSEC compared to B80E20 in all test conditions. At the 8.1 kW load, B76E19DEE5 presented the lowest BSEC with respect to the other fuels tested. These results can be attributed to the improvement of the burning process due to the oxygen content and the elevation of the CN of the blend with DEE [16–18].

3.1.3. Engine Efficiency

Engine efficiency may be defined as the useful energy output of the engine as a function of fuel heat energy. As shown in Figure 4, higher loads present increased engine efficiency as a result of a more adequate air/fuel ratio and enhanced mechanical efficiency [20].

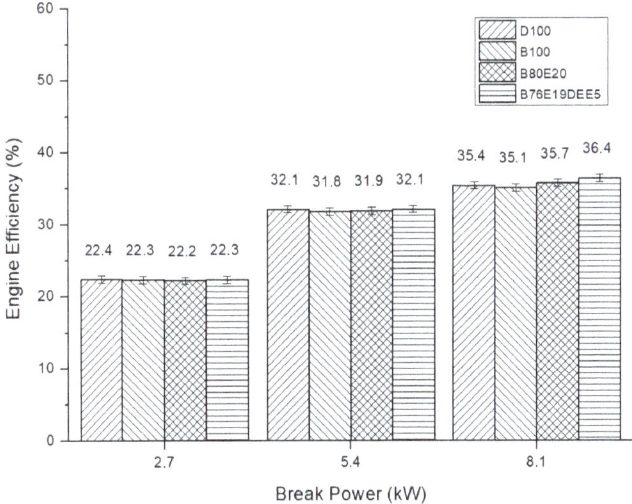

Figure 4. The variation of engine efficiency with respect to the engine load.

A slight reduction of about 0.7% on average, under all loads, was observed in engine efficiency while using B100 with respect to D100. These variations can be attributed to differences in viscosity and density, which influence atomization and decrease combustion efficiency [21]. This can be confirmed by a more significant reduction in BSFC in comparison to that observed for the LHVs when B100 replaces D100. The results are consistent with some studies in the literature [22–25] despite other studies show increased efficiency while the engine operates with blends of biodiesel and diesel. It can be discussed that when biodiesel raw material presents low viscosity, the oxygen in the biodiesel molecule and the higher cetane number than diesel can improve the combustion characteristic [26–28]. On the other hand, addition of ethanol to the blend decreases viscosity, LHV, and CN in comparison with B100, although oxygen content increases. When related to B100, the blend B80E20 presents an increase in efficiency of 0.4% and 1.6% at 5.4 kW and 8.1 kW loads, respectively. The lower viscosity of B80E20 improves atomization and provides better evaporation and air-fuel mixing, hence improving combustion efficiency. In addition, higher oxygen content causes faster combustion [7,25,29]. At 2.7 kW, similar efficiencies values were observed. Introduction of diethyl ether (B76E19DEE5) revealed a slight increase of 0.5% in engine efficiency under 2.7 kW and 5.4 kW of engine loads in comparison with B80E20 fuel, whereas under 8.1 kW the increase was about 2%. DEE addition to the blend promotes a reduction in viscosity, which facilitates fuel atomization, and due to the higher CN of diethyl ether, it decreases the ignition delay [18].

3.2. Emissions

3.2.1. NO_x Emissions

NO_x emissions for the studied blends compared to D100 and B100 are shown in Figure 5. It has been widely discussed that as loads increases, NO_x exhaust emission increases due to higher temperatures in the combustion chamber that increases the NO_x formation by thermal mechanism [30–32].

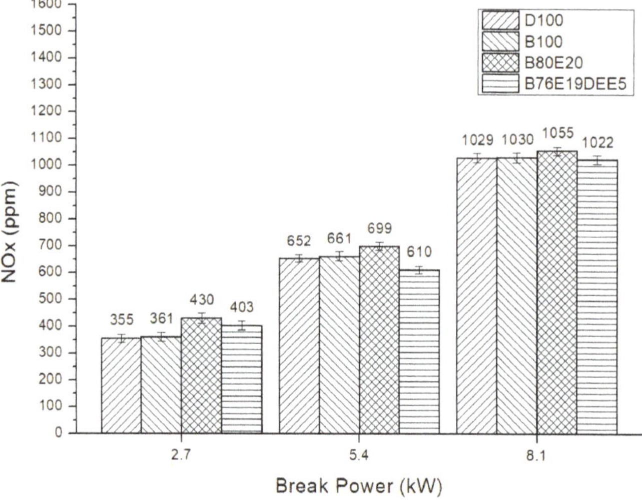

Figure 5. The variation of NO_x emissions with respect to the engine load.

B100 showed minor increases in NO_x emissions in comparison with fossil diesel. This increase can be explained primarily by the oxygen content in the biodiesel molecule, which increases oxygen availability and combustion temperatures [33]. Besides, the higher bulk modulus of biodiesel can advance its injection increasing the preparation time of the fuel-air mixture, which increases combustion pressure and temperature, thus, contributing to NO_x formation [33].

B80E20 fuel resulted in an increase in NO_x emissions compared to the other tested fuels. These results can be attributed to the lower CN, which prolongs the premix formation time and increases heat release ratio (HRR) in the initial phase of combustion, thus higher temperatures and NO_x emission are obtained [34,35]. Furthermore, the high oxygen content of ethanol (Table 2) contributes to NO_x formation [8]. The high oxygen content of the fuel has a positive effect on the combustion since the oxygen presented in the fuel is more active when compared to the molecular oxygen contained in the air [36]. Even though higher oxygen content produces greater combustion efficiency, higher temperatures increase NO_x formation [37,38].

The use of DEE (B76E19DEE5) revealed a significant reduction in NO_x emission. In comparison with the B80E20 blend, the reduction was approximately 6%, 13%, and 3%, at 2.7 kW, 5.4 kW, and 8.1 kW engine loads, respectively. Besides, under 5.4 kW and 8.1 kW, the B76E19DEE5 blend has shown the lowest NO_x emission among all fuels tested. These results can be attributed to two main factors that are mainly related to fuel characteristics. Firstly, the higher CN, which decreases the ignition delay and the premix formation time, thus reducing the combustion pressure and temperature peaks. Furthermore, the enthalpy of vaporization of the blend is also elevated, which reduces combustion temperature, thus, reducing NO_x emissions [8].

3.2.2. CO Emissions

Figure 6 shows the CO emission for the fuels tested. CO emissions increased for higher engine load due to lower air-fuel ratios presented in the combustion phase [37].

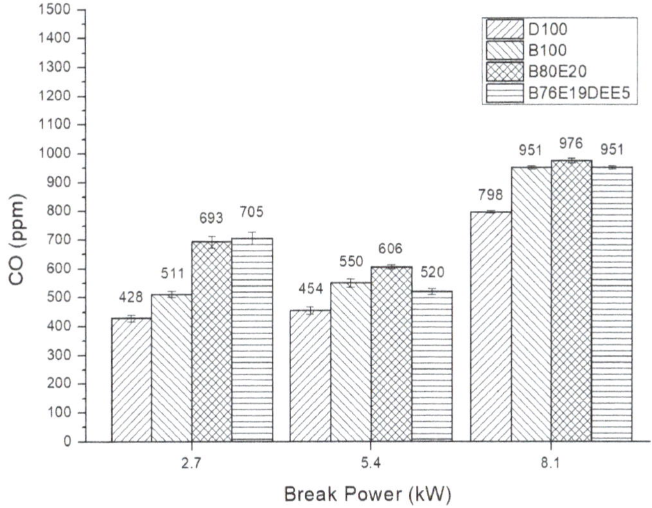

Figure 6. The variation of CO emissions with respect to the engine load.

The use of B100 increases CO emissions compared to D100 in about 20%. This may be attributed to the lower air-fuel ratio and poor atomization that results due to biodiesel higher viscosity [37,38]. Some studies also indicate a reduction in CO emissions when the engine operates with biodiesel [27,31] as a result of the higher oxygen concentration. Blending ethanol with biodiesel (B80E20) considerably increased CO emission in comparison with B100 under all loads up to 35%. This is attributed to the reduction in combustion temperature due to the high enthalpy of vaporization of ethanol [8]. This result is consistent with the result found in Çelik et al. [7]. The addition of DEE to biodiesel-ethanol blend produced a slight decrease in CO emission in comparison with B80E20, particularly under medium and high engine load. The higher cetane number and increased oxygen content of the B76E19DEE5 blend improve combustion efficiencies and decrease CO emission [39–41]. At 2.7 kW, however, a slight increase was verified and has been attributed due to the decrease in combustion temperature as a result of both alcohol and DEE cooling effect into the blend.

3.2.3. THC Emissions

Total hydrocarbons emissions of the three engine operating loads (Figure 7) with blends of B80E20 and B76E19DEE5 were higher than those of D100 and B100.

The increase of THC emissions when ethanol is blended with diesel and biodiesel has been previously reported in the literature [35,42]. Emissions decreased with increasing engine loads for all tested fuels. This is due to the higher combustion temperature at higher loads that enhances combustion efficiency and thus reduces unburnt THC [43].

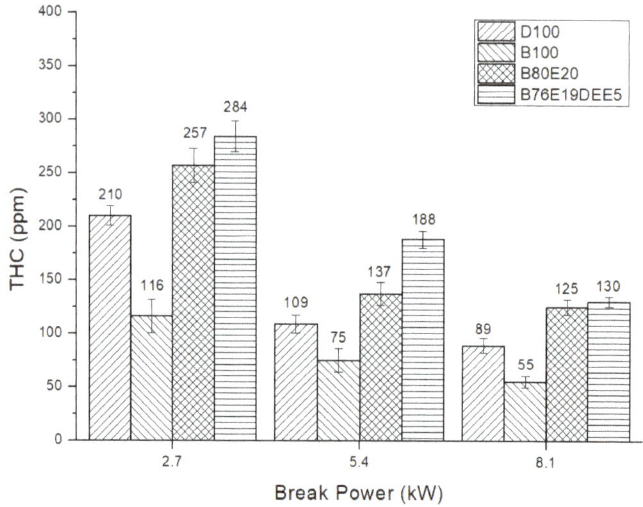

Figure 7. The variation of THC emissions with respect to the engine load.

The use of B100 shows significant reductions in THC emissions by 45%, 31%, and 38% at 2.7 kW, 5.4 kW, and 8.1 kW loads, respectively, in comparison with D100. Among the factors responsible for the decrease in hydrocarbon emission is the higher CN and oxygen content of B100 since the oxygen contained in the fuel provides cleaner and more complete combustion related to D100. Besides, biodiesel combustion starts earlier in the combustion chamber due to its higher bulk modulus providing more time for mixture formation [44]. Similar results were previously reported in the literature [45,46]. B80E20 fuel shows an increase in THC emission compared to B100 and D100. Although the oxygen content in ethanol could increase the combustion quality, other properties have the opposite influence, such as lower CN, lower LHV, and higher enthalpy of vaporization. The higher enthalpy of vaporization reduces combustion temperature, which influences the oxidation rate and THC formation [8,47]. The lower CN increases the ignition delay and decreases the total combustion time, which can increase the THC formation [45]. A reduction in LHV due to ethanol addition leads a higher amount of fuel injected per cycle, which also may favor THC formation [8].

The addition of diethyl ether to biodiesel-ethanol blend (B76E19DEE5) increased THC emission in comparison with the B80E20 blend. This result may be attributed to the high enthalpy of vaporization and high volatility of the blend when DEE was introduced, which reduces combustion temperature related to B80E20 [35,48].

4. Conclusions

Emissions, fuel consumption, and efficiency of a DI mechanical diesel engine coupled to a hydraulic dynamometer were measured using fossil diesel, biodiesel, biodiesel-ethanol (B80E20), and biodiesel-ethanol-diethyl ether (B76E19DEE5). Although BSFC increased with the use of biofuels due to the reduction in LHV, engine efficiencies increased with the use of B80E20 and B76E19DEE5, especially at medium and high loads. The higher oxygen content was appointed as the main reason for that improvement. In the case of the B76E19DEE5 fuel, the CN increased in comparison with B80E20, which further enhance the engine efficiency. Considering BSEC, the similarity of results shows that the performance of the proposed fuel blend was similar with the conventional diesel fuel, used in the original engine design. However, at the 8.1 kW load, B76E19DEE5 presented the lowest BSEC in relation to the other fuels tested. This lower BSEC was consistent with the highest engine efficiency in the same condition.

The use of B100 increased NO_x and CO emissions compared to D100 (1% and 20%, respectively, on average), which was attributed to the higher viscosity of biodiesel. On the other hand, THC emissions decreased with the use of B100 due to the oxygen content in biodiesel, which provides a cleaner and more complete combustion when compared with D100. Addition of ethanol (B80E20) showed an increase in NO_x emissions compared to B100 as well as compared to the other fuels. The results were attributed to the higher oxygen content and lower CN in the mixture. Considerable increases in CO and THC emissions were attributed to the high enthalpy of vaporization and lower LHV of ethanol. Finally, addition of DEE to biodiesel-ethanol blend (B76E19DEE5) generated significant reductions in NO_x emissions. In comparison with B80E20 fuel, the reductions (3% to 13%) were attributed to the higher CN and the higher enthalpy of vaporization of the blend. At 5.4 kW and 8.1 kW loads, NO_x emissions were the lowest among all tested fuels. Regarding CO emissions, the results showed reductions (3% to 14%) in 5.4 kW and 8.1 kW loads, respectively, compared to B80E20 fuel. It was attributed to the higher CN and oxygen content of the blend. At 2.7 kW load, however, only a slight increase was verified. The THC emissions raised in comparison with B80E20 fuel up to 37% at the high engine load. The results were attributed to the higher enthalpy of vaporization and by the elevated oxygen content of the B76E19DEE5 blend. No problems with engine stability were noticed using the blends proposed. This could be verified by low variation in the engine speed and at the values of torque acquisition.

It was clear in this work the positive effects of adding DEE in moderate concentration to the ethanol biodiesel mixture, in which there were slight increases in engine efficiency and reductions in NO_x emissions.

Author Contributions: Conceptualization, M.C., J.S. and V.F.; Data curation, M.C.; Formal analysis, M.C., J.S., V.F., F.T., J.M. and E.T.; Funding acquisition, E.T.; Investigation, M.C.; Methodology, M.C., J.S., J.M. and V.F.; Project administration, M.C.; Resources, M.C. and E.T.; Supervision, J.M., V.F. and E.T.; Validation, M.C. and V.F.; Writing, M.C., J.M., V.F. and F.T.; Writing—review & editing, M.C., F.T., J.S., V.F., J.M. and E.T. All authors have read and agreed to the published version of the manuscript.

Funding: This work had the financial support of Energy Laboratory of Bahia Federal University (UFBA) and CAPES (Coordination for the Improvement of Higher Education Personnel).

Acknowledgments: Energy Laboratory of Bahia Federal University (UFBA), Postgraduate Program in Industrial Engineering (PEI/ UFBA), and CAPES (Coordination for the Improvement of Higher Education Personnel). Authors also acknowledge PETROBAHIA for providing fuels to the work: diesel pure, biodiesel, and anhydrous ethanol fuels.

Conflicts of Interest: The authors declare no conflict of interest.

References

1. Mahbub, N.; Gemechu, E.; Zhang, H.; Kumar, A. The life cycle greenhouse gas emission benefits from alternative uses of biofuel coproducts. *Sustain. Energy Technol. Assess.* **2019**, *34*, 173–186. [CrossRef]
2. Ospina, G.; Selim, M.Y.E.; al Omari, S.A.B.; Ali, M.I.H.; Hussien, A.M.M. Engine roughness and exhaust emissions of a diesel engine fueled with three biofuels. *Renew. Energy* **2019**, *134*, 1465–1472. [CrossRef]
3. Khoobbakht, G.; Najafi, G.; Karimi, M.; Akram, A. Optimization of operating factors and blended levels of diesel, biodiesel and ethanol fuels to minimize exhaust emissions of diesel engine using response surface methodology. *Appl. Therm. Eng.* **2016**, *99*, 1006–1017. [CrossRef]
4. Noor, C.W.M.; Noor, M.M.; Mamat, R. Biodiesel as alternative fuel for marine diesel engine applications: A review. Renew. *Sustain. Energy Rev.* **2017**, *94*, 127–142. [CrossRef]
5. Garzón, N.N.A.; Oliveira, A.A.; Hartmann, M.R.; Bazzo, E. Experimental and thermodynamic analysis of a compression ignition engine operating with straight soybean oil. *J. Braz. Soc. Mech. Sci. Eng.* **2015**, *37*, 1467–1478. [CrossRef]
6. Ferreira, V.P.; Martins, J.; Torres, E.A.; Pepe, I.; De Souza, J.M.S. Performance and emissions analysis of additional ethanol injection on a diesel engine powered with a blend of diesel-biodiesel. *Energy Sustain. Dev.* **2013**, *17*, 649–657. [CrossRef]

7. Çelik, M.; Örs, İ.; Bayindirli, C.; Demiralp, M. Experimental investigation of impact of addition of bioethanol in different biodiesels, on performance, combustion and emission characteristics. *J. Mech. Sci. Technol.* **2017**, *31*, 5581–5592. [CrossRef]
8. Venu, H.; Madhavan, V. Influence of diethyl ether (DEE) addition in ethanol-biodiesel-diesel (EBD) and methanol-biodiesel-diesel (MBD) blends in a diesel engine. *Fuel* **2017**, *189*, 377–390. [CrossRef]
9. Al-Esawi, N.; Al Qubeissi, M.; Kolodnytska, R. The impact of biodiesel fuel on ethanol/diesel blends. *Energies* **2019**, *12*, 1804. [CrossRef]
10. Al Qubeissi, M.; Al-Esawi, N.; Sazhin, S.S.; Ghaleeh, M. Ethanol/gasoline droplet heating and evaporation: Effects of fuel blends and ambient conditions. *Energy Fuels* **2018**, *32*, 6498–6506. [CrossRef]
11. Al-Esawi, N.; Al Qubeissi, M.; Whitaker, R.; Sazhin, S.S. Blended E85–Diesel fuel droplet heating and evaporation. *Energy Fuels* **2019**, *33*, 2477–2488. [CrossRef]
12. Tutak, W.; Jamrozik, A.; Pyrc, M.; Sobiepański, M. A comparative study of co-combustion process of diesel-ethanol and biodiesel-ethanol blends in the direct injection diesel engine. *Appl. Therm. Eng.* **2017**, *117*, 155–163. [CrossRef]
13. Yilmaz, N. Performance and emission characteristics of a diesel engine fuelled with biodiesel-ethanol and biodiesel-methanol blends at elevated air temperatures. *Fuel* **2012**, *94*, 440–443. [CrossRef]
14. Prbakaran, B.; Viswanathan, D. Experimental investigation of effects of addition of ethanol to bio-diesel on performance, combustion and emission characteristics in CI engine. *Alex. Eng. J.* **2018**, *57*, 383–389. [CrossRef]
15. Kandasamy, S.K.; Selvaraj, A.S.; Rajagopal, T.K.R. Experimental investigations of ethanol blended biodiesel fuel on automotive diesel engine performance, emission and durability characteristics. *Renew Energy* **2019**, *141*, 411–419. [CrossRef]
16. Ibrahim, A. Investigating the effect of using diethyl ether as a fuel additive on diesel engine performance and combustion. *Appl. Therm. Eng.* **2016**, *107*, 853–862. [CrossRef]
17. Lee, S.; Kim, T.Y. Performance and emission characteristics of a DI diesel engine operated with diesel/DEE blended fuel. *Appl. Therm. Eng.* **2017**, *121*, 454–461. [CrossRef]
18. Qi, D.H.; Chen, H.; Geng, L.M.; Bian, Y.Z. Effect of diethyl ether and ethanol additives on the combustion and emission characteristics of biodiesel-diesel blended fuel engine. *Renew Energy* **2011**, *36*, 1252–1258. [CrossRef]
19. Jeevanantham, A.K.; Nanthagopal, K.; Ashok, B.; Al-Muhtaseb, A.H.; Thiyagarajan, S.; Geo, V.E.; Ong, H.C.; Samuel, K.J.; Chyuan, O.H. Impact of addition of two ether additives with high speed diesel- Calophyllum Inophyllum biodiesel blends on NO_x reduction in CI engine. *Energy* **2019**, *185*, 39–54. [CrossRef]
20. Heywood, J.B. *Internal Combustion Engine Fundamentals*; McGraw-Hill: New York, NY, USA, 1988.
21. Du, E.; Cai, L.; Huang, K.; Tang, H.; Xu, X.; Tao, R. Reducing viscosity to promote biodiesel for energy security and improve combustion efficiency. *Fuel* **2018**, *211*, 194–196. [CrossRef]
22. Hulwan, D.B.; Joshi, S.V. Performance, emission and combustion characteristic of a multicylinder DI diesel engine running on diesel-ethanol-biodiesel blends of high ethanol content. *Appl. Energy* **2011**, *88*, 5042–5055. [CrossRef]
23. Veinblat, M.; Baibikov, V.; Katoshevski, D.; Wiesman, Z.; Tartakovsky, L. Impact of various blends of linseed oil-derived biodiesel on combustion and particle emissions of a compression ignition engine—A comparison with diesel and soybean fuels. *Energy Convers. Manag.* **2018**, *178*, 178–189. [CrossRef]
24. Duda, K.; Wierzbicki, S.; Śmieja, M.; Mikulski, M. Comparison of performance and emissions of a CRDI diesel engine fuelled with biodiesel of different origin. *Fuel* **2018**, *212*, 202–222. [CrossRef]
25. Elkelawy, M.; Bastawissi, H.A.-E.; Esmaeil, K.K.; Radwan, A.M.; Panchal, H.; Sadasivuni, K.K.; Ponnamma, D.; Walvekar, R. Experimental studies on the biodiesel production parameters optimization of sunflower and soybean oil mixture and DI engine combustion, performance, and emission analysis fueled with diesel/biodiesel blends. *Fuel* **2019**, *255*, 115791. [CrossRef]
26. Seraç, M.R.; Aydın, S.; Yılmaz, A.; Şevik, S. Evaluation of comparative combustion, performance, and emission of soybean-based alternative biodiesel fuel blends in a CI engine. *Renew. Energy* **2020**, *148*, 1065–1073. [CrossRef]
27. Lapuerta, M.; Armas, O.; Rodríguez-Fernández, J. Effect of biodiesel fuels on diesel engine emissions. *Prog. Energy Combust. Sci.* **2008**, *34*, 198–223. [CrossRef]
28. Palash, S.M.; Masjuki, H.H.; Kalam, M.a.; Masum, B.M.; Sanjid, A.; Abedin, M.J. State of the art of NO_x mitigation technologies and their effect on the performance and emission characteristics of biodiesel-fueled Compression Ignition engines. *Energy Convers. Manag.* **2013**, *76*, 400–420. [CrossRef]

29. Wei, L.; Cheung, C.S.; Ning, Z. Effects of biodiesel-ethanol and biodiesel-butanol blends on the combustion, performance and emissions of a diesel engine. *Energy* **2018**, *155*, 957–970. [CrossRef]
30. Abed, K.A.; Gad, M.S.; el Morsi, A.K.; Sayed, M.M.; Elyazeed, S.A. Effect of biodiesel fuels on diesel engine emissions. *Egypt. J. Pet.* **2019**, *28*, 183–188. [CrossRef]
31. Song, J.; Zello, V.; Boehman, A.L.; Waller, F.J. Comparison of the impact of intake oxygen enrichment and fuel oxygenation on diesel combustion and emissions. *Energy Fuels* **2004**, *18*, 1282–1290. [CrossRef]
32. Rahman, M.A.; Aziz, M.A.; Ruhul, A.M.; Rashid, M.M. Biodiesel production process optimization from Spirulina maxima microalgae and performance investigation in a diesel engine. *J. Mech. Sci. Technol.* **2017**, *31*, 3025–3033. [CrossRef]
33. Rajkumar, S.; Thangaraja, J. Effect of biodiesel, biodiesel binary blends, hydrogenated biodiesel and injection parameters on NO_x and soot emissions in a turbocharged diesel engine. *Fuel* **2019**, *240*, 101–118. [CrossRef]
34. Aydin, H.; Ilkiliç, C. Effect of ethanol blending with biodiesel on engine performance and exhaust emissions in a CI engine. *Appl. Therm. Eng.* **2010**, *30*, 1199–1204. [CrossRef]
35. Jamrozik, A. The effect of the alcohol content in the fuel mixture on the performance and emissions of a direct injection diesel engine fueled with diesel-methanol and diesel-ethanol blends. *Energy Convers. Manag.* **2017**, *148*, 461–476. [CrossRef]
36. Gharehghani, A.; Mirsalim, M.; Hosseini, R. Effects of waste fish oil biodiesel on diesel engine combustion characteristics and emission. *Renew. Energy* **2017**, *101*, 930–936. [CrossRef]
37. Xue, J.; Grift, T.E.; Hansen, A.C. Effect of biodiesel on engine performances and emissions. *Renew. Sustain. Energy Rev.* **2011**, *15*, 1098–1116. [CrossRef]
38. Santos, T.B.; Ferreira, V.P.; Torres, E.A.; Julio, A.M.; Ordonez, J.C. Energy analysis and exhaust emissions of a stationary engine fueled with diesel—Biodiesel blends at variable loads. *J. Braz. Soc. Mech. Sci. Eng.* **2017**, *39*, 3237–3247. [CrossRef]
39. Paul, A.; Bose, P.K.; Panua, R.; Debroy, D. Study of performance and emission characteristics of a single cylinder CI engine using diethyl ether and ethanol blends. *J. Energy Inst.* **2015**, *88*, 1–10. [CrossRef]
40. Beatrice, C.; Avolio, G.; Del Giacomo, N.; Guido, C.; Lazzaro, M. The effect of "Clean and Cold" EGR on the improvement of low temperature combustion performance in a single cylinder research diesel engine. *SAE Tech. Pap.* **2008**. 2008-01-650.
41. Sukjit, E.; Herreros, J.M.; Dearn, K.D.; García-Contreras, R.; Tsolakis, A. The effect of the addition of individual methyl esters on the combustion and emissions of ethanol and butanol -diesel blends. *Energy* **2012**, *42*, 364–374. [CrossRef]
42. Júnior, L.C.S.S.; Ferreira, V.P.; da Silva, J.A.M.; Torres, E.A.; Pepe, I.M. Oxidized biodiesel as a cetane improver for diesel–biodiesel–ethanol mixtures in a vehicle engine. *J. Braz. Soc. Mech. Sci. Eng.* **2018**, *40*, 79.
43. Shamun, S.; Belgiorno, G.; di Blasio, G.; Beatrice, C.; Tunér, M.; Tunestal, P. Performance and emissions of diesel-biodiesel-ethanol blends in a light duty compression ignition engine. *Appl. Therm. Eng.* **2018**, *145*, 444–452. [CrossRef]
44. Caresana, F. Impact of biodiesel bulk modulus on injection pressure and injection timing the effect of residual pressure. *Fuel* **2011**, *90*, 477–485. [CrossRef]
45. Lahane, S.; Subramanian, K.A. Effect of different percentages of biodiesel–diesel blends on injection, spray, combustion, performance, and emission characteristics of a diesel engine. *Fuel* **2015**, *139*, 537–545. [CrossRef]
46. An, H.; Yang, W.M.; Li, J.; Zhou, D.Z. Modeling analysis of urea direct injection on the NO_x emission reduction of biodiesel fueled diesel engines. *Energy Convers. Manag.* **2015**, *101*, 442–449. [CrossRef]
47. Çelebi, Y.; Aydın, H. An overview on the light alcohol fuels in diesel engines. *Fuel* **2019**, *236*, 890–911. [CrossRef]
48. Rakopoulos, D.C.; Rakopoulos, C.D.; Giakoumis, E.G.; Papagiannakis, R.G.; Kyritsis, D.C. Influence of properties of various common bio-fuels on the combustion and emission characteristics of high-speed DI (direct injection) diesel engine: Vegetable oil, bio-diesel, ethanol, n-butanol, diethyl ether. *Energy* **2014**, *73*, 354–366. [CrossRef]

Article

Improving Fuel Economy and Engine Performance through Gasoline Fuel Octane Rating

José Rodríguez-Fernández [1,*], Ángel Ramos [1], Javier Barba [1], Dolores Cárdenas [2] and Jesús Delgado [2]

[1] Escuela Técnica Superior de Ingeniería Industrial, University of Castilla—La Mancha, 13071 Ciudad Real, Castilla–La Mancha, Spain; angel.ramos@uclm.es (Á.R.); Javier.Barba@uclm.es (J.B.)

[2] Repsol Technology Lab, 28935 Móstoles, Madrid, Spain; mdcardenasa@repsol.com (D.C.); jdelgadod@repsol.com (J.D.)

* Correspondence: jose.rfernandez@uclm.es

Received: 19 June 2020; Accepted: 2 July 2020; Published: 7 July 2020

Abstract: The octane number is a measure of the resistance of gasoline fuels to auto-ignition. Therefore, high octane numbers reduce the engine knocking risk, leading to higher compression threshold and, consequently, higher engine efficiencies. This allows higher compression ratios to be considered during the engine design stage. Current spark-ignited (SI) engines use knock sensors to protect the engine from knocking, usually adapting the operation parameters (boost pressure, spark timing, lambda). Moreover, some engines can move the settings towards optimized parameters if knock is not detected, leading to higher performance and fuel economy. In this work, three gasolines with different octane ratings (95, 98 and 100 RON (research octane number)) were fueled in a high-performance vehicle. Tests were performed in a chassis dyno at controlled ambient conditions, including a driving sequence composed of full-load accelerations and two steady-state modes. Vehicle power significantly increased with the octane rating of the fuel, thus decreasing the time needed for acceleration. Moreover, the specific fuel consumption decreased as the octane rating increased, proving that the fuel can take an active part in reducing greenhouse gas emissions. The boost pressure, which increased with the octane number, was identified as the main factor, whereas the ignition advance was the second relevant factor.

Keywords: octane number; knocking; spark-ignition; performance; knock sensor; fuel economy; vehicle acceleration

1. Introduction

The European Union (EU) has recently committed to achieving carbon neutrality by 2050 [1]. This goal necessarily involves diminishing CO_2 emissions in the transport sector, responsible for 27% of total European greenhouse gas (GHG) emissions [2]. In this sector, well-to-wheel (WtW) analyses estimate the GHG emissions associated with the fuel, inventorying emissions in feedstock-related (or primary fuel-related) stages, fuel-related stages and final fuel use in the vehicle. The last step is known as tank-to-wheel emissions, which can be mitigated through fuel formulation and more efficient vehicle and engine technologies. For this reason, present and future research on engines pursues increasing the efficiency [3,4], thus achieving better fuel economy and lower CO_2 emissions.

In the case of spark-ignited (SI) engines, direct injection (DI) and, more recently, downsizing, turbocharging and high compression ratios are the main working areas for increasing their efficiency. However, all these approaches are constrained by the appearance of abnormal combustion regimes. The fuel itself is fundamental for overcoming this limitation, and fuel manufacturers can accompany GHG reduction by developing high-octane gasolines that could be properly exploited by current SI engine technologies. This way, fuel and engine technologies interact synergistically, with the fuel enabling the engine to work on more efficient conditions. CONCAWE (Environmental Science for European Refining, an association of companies that operate petroleum refineries in the EU) [5] recently modelled and tested the potential of high-octane gasolines to enhance the efficiency of downsized, high-compression ratio SI engines (whose share in the market is expected to grow according to the current trends). The highest-octane fuel (102 RON—Research Octane Number) improved fuel economy around 4% in driving cycles, compared to 95 RON.

The anti-knocking tendency of a gasoline has been traditionally described by the octane numbers, the research octane number (RON) and the motor octane number (MON, measured under more stressed testing conditions), both included in EN 228 Standard for unleaded petrol quality. However, some studies show that in current SI engines MON is no longer a good indicator and higher MON can indeed be unfavourable to engine performance [6]. In the EU, gasoline vehicles must operate safely with regular 95 RON fuel, but there is a range of vehicles that can take advantage of higher-octane gasolines to increase the efficiency when running under adverse conditions (high ambient temperature, full-load accelerations) [5].

The most common undesired combustion phenomenon is knock [7] (although cases of benign knock that leads to higher efficiency have been reported [8]), but other unwanted regimes are super-knock and pre-ignition [7]. Knock, or knocking, occurs when the fuel-air mixture in the unburnt gas zone auto ignites ahead of the flame front. The presence of hot spots and higher temperature and pressure (the above-referred trends in SI engines contribute to these) in the combustion chamber makes knock more probable, as well as low speed and high load conditions. With the strengthening of turbocharging and downsizing, super-knock (a knock phenomenon at higher intensity than usual) has been documented and explained based on developing detonations [9]. Pre-ignition is caused by hot spots inside the combustion chamber, such as deposits in the spark plugs and valves [10]. However, in modern DI engines the most probable cause for pre-ignition is the accumulation of lubricant droplets in the chamber when the fuel spray impinges the cylinder walls [11]. This and other causes have been reviewed in [12]. Pre-ignition typically occurs in the compression stroke, earlier than the command from the spark plug, and thus decreases the efficiency and the power. Pre-ignition makes more likely the appearance of super-knock [13].

A few transient episodes of abnormal combustion are not dangerous [3]. However, prolonged operation deteriorates the engine performance, damages the engine (pistons and electrodes, mainly) [14], thus affecting engine reliability and durability, and produces noise and driver annoyance. Hence, detecting and suppressing knock is essential.

Although there are other methods (based on noise measurements [15]), knock can be mainly detected with a pressure sensor located inside the cylinder (or several cylinders) or using a vibration sensor (called "knock" sensor). The first is typical in research and development applications, sometimes in combination with optical techniques [16], as this provides a complete understanding of the combustion. When the amplitude of the pressure fluctuations exceeds a threshold, the engine is knocking [4]. By contrast, most commercial vehicles are equipped with knock sensors, which are accelerometers that sense the vibrations that knocking and uneven combustion in the cylinders cause in the engine block. The output signal of the sensor is informed to the engine control unit (ECU) to decide a corrective action that returns the engine to regular combustion, despite this implying a power reduction or/and fuel economy penalization.

The most acknowledged corrective action consists in retarding the ignition [17,18]. Other measures include changing the intake valve timing to reduce the effective compression ratio [19], enriching the fuel–air mixture [20] (for the excess fuel to act as a charge cooler), using EGR [21], specially cooled EGR [22] (because EGR prolongs the ignition delay time of the unburnt gas, cooling the intake charge [23]) or limiting the boost pressure [24], among others. Today, knock sensors are essential not only for ensuring durability but also for improving the engine efficiency and fuel economy and for reducing emissions.

In current SI engines, for a given operating point and fuel, advancing the spark timing increases the engine efficiency but also the knock probability [4]. The spark advance for which the knock intensity is unacceptable is called knock-limited spark advance (KLSA) [25], and this is usually achieved earlier than the spark advance for maximum efficiency (maximum brake torque, MBT). The latest trends in SI engines (turbocharging, higher compression ratios) have indeed accentuated this difference, thus enlarging the potential of high-octane gasolines to increase the efficiency.

Some works evaluate the effects of high-octane gasolines in SI engines. In [26], the authors noticed that a higher octane gasoline increased torque and power only in an engine with a knock control system. Stradling et al. [6] tested acceleration sequences with different octane rating gasolines (including ethers and alcohols) in two passenger cars equipped with knock sensors. The acceleration time decreased (power increased) and the energy consumption deteriorated with the octane number, this trend being more accentuated in the low-octane range. Moreover, both variables fitted better with RON than MON (higher MON was detrimental, actually). Although the effect of the octane number on the emissions was also explored, emissions are affected by the fuel molecular structure as well. Shuai et al. [27] tested five vehicles and found different sensitivities to the fuel. The effect of the octane number was more notable at high speed and load. On average, they reported 1% better fuel economy per unit of RON.

In the present work, three gasoline fuels meeting all requirements of EN 228 Standard, with regular (95 RON), middle (98 RON) and high octane (100 RON), have been tested in a passenger car at full load conditions in both accelerations and steady-state modes; 100 RON gasoline was prepared with a N-aquil substituted aniline as octane booster, which has not been tested in high-performance commercial vehicles yet. The results include not only fuel consumption, thermal efficiency and CO_2 emissions, but also the rest of the gaseous-regulated emissions (CO, total hydrocarbon (THC), NOx) for a complete evaluation of the fuels. The study aligns with the trend in the fuel market towards high-octane gasolines to support the introduction and consolidation of higher efficiency engines. In fact, the recent update of the Worldwide Fuel Charter [28] released by the main automobile associations introduces a new gasoline category, characterized by high octane numbers mainly, intended for the most stringent markets in terms of CO_2 targets.

2. Materials and Methods

2.1. Experimental Setup

The study was carried out on a two-wheel drive (2WD) chassis dynamometer Schenk for light-duty vehicles, which is located inside a climatic chamber (Figure 1). It is equipped with a single roller (159.5 cm diameter, 168 kW nominal power) which simulates the rolling and the aerodynamic resistances, as well as the equivalent vehicle inertia. The pressure and temperature sensors required are located according to [29]. The blower (placed in front of the vehicle) is used to produce cooling wind at the simulated vehicle velocity. Inside the climatic chamber, the ambient temperature can be regulated from −20 °C to 40 °C.

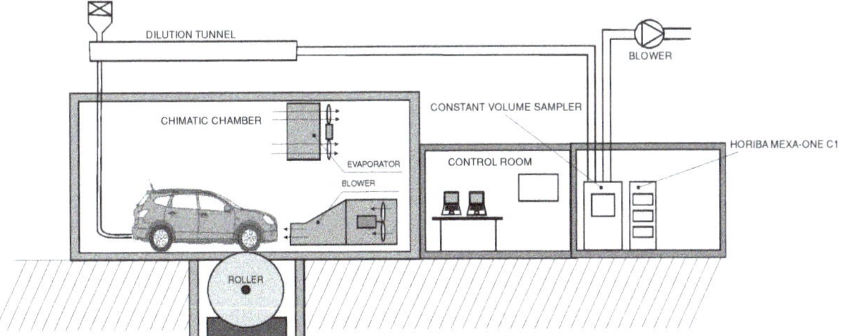

Figure 1. Scheme of the experimental setup.

The total exhaust gas flow rate emitted by the vehicle is diluted with filtered atmospheric air in a total dilution tunnel Horiba DLT-7040. The pollutant emissions were measured both continuously (time-resolved) and after a small diluted gas sample was collected in bags. Gaseous emissions were measured with a Horiba MEXA-ONE C1 analyser. It is equipped with separate modules for the measurement of nitrogen oxides, carbon dioxide, carbon monoxide and total hydrocarbons. Sampling probes and transport lines from the measurement point (dilution tunnel) were heated at 191 °C to avoid condensation of hydrocarbons. The specifications of the equipment are shown in Table 1. All measurements were previously synchronised, since the response time of each module and the delay time in the transport lines are different.

Table 1. Specifications of Horiba MEXA-ONE C1.

Carbon Monoxide Module	
Measurement method	Non-dispersive infrared (NDIR)
Range	0–5000 ppm
Accuracy	3% (of measured value)
Carbon Dioxide Module	
Measurement method	Non-dispersive infrared (NDIR)
Range	0–1000 ppm
Accuracy	3.1% (of measured value)
Nitrogen Oxides Module	
Measurement method	Chemiluminescence (CLD), vacuum
Range	0–5000 ppm
Accuracy	2.7% (of measured value)
Total hydrocarbon module	
Measurement method	Flame ionization detector (FID)
Range	0–60,000 ppm
Accuracy	2.7% (of measured value)

Regarding fuel consumption, this was calculated based on the carbon balance (CO_2, CO and THC) with the method proposed by Directive 1151/2017 of the European Commission [29]. An Opel Corsa OPC equipped with a spark-ignition indirect injection (SI-IDI) engine was selected for the tests (Table 2). This is a commercial high-performance vehicle with maximum power within the limits of the experimental facility. The vehicle was not modified prior to gasoline testing in order to identify real advantages on vehicles available in the current market, without considering future design improvements that are expected and have been discussed in other works. The engine control strategy of the vehicle was not modified from the original one; therefore, the results are expected to reproduce well the on-road performance and emissions. Several variables were measured and recorded through OBD (on-board diagnostics), such as vehicle velocity, accelerator pedal position, boost pressure, lambda sensor or spark timing, among others, in order to evaluate the vehicle performance and support the trends observed.

2.2. Fuels

The samples used in this study present different antiknock properties and satisfy the specifications for winter gasoline EN-228 for a maximum quantity of oxygen of 2.7% (E5). Two of them are commercial type, sampled from Company Service Stations, representing the grades distributed in Spain (95 RON and 98 RON). The third gasoline is a modification of 98 RON commercial gasoline by using an octane booster, with a target of 100 RON. Octane booster is an N-aquil substituted aniline with high efficiency to increase RON without relevant changes in the rest of the properties.

Table 2. Specifications of Opel Corsa OPC.

Emission Regulation	Euro 6b
Engine type	SI-IDI, turbocharged, intercooler
Compression ratio	8.8:1
Power @ 5800 rpm	152 kW
Torque @ 1.900–5.800 rpm	280 Nm
Displacement	1.598 cm^3
Bore	79 mm
Stroke	81.5 mm
Gearbox	Manual
Valves per cylinder	4
Octane number (RON) recommended [1]	100
Total gear ratio (km/h each 1000 rpm)	
1st	7.2
2nd	12.7
3rd	20.3
4th	28.6
5th	35.7
6th	44.6

[1] As specified by the vehicle manufacturer.

Table 3 shows the key properties of the different evaluated samples. The heating values of all fuels were similar. The small differences were within the reproducibility of the testing (ASTM D240). The selection of the samples answered to the need to identify the advantages of using high-octane gasolines, compared to standard gasoline. Performance, consumption and emissions will define the added value to the customer and the potential of each product. The 100 RON gasoline is a new high-octane product in Spanish market. Therefore, a rigorous measurement of improvements with respect to existing products was the key objective of this study.

Table 3. Properties of tested gasolines.

Test	Method	Units	95 RON	98 RON	100 RON
Research Octane Number	ASTM* D 2699-18a	–	96.1	98.1	99.7
Motor Octane Number	ASTM D 2700-18a	–	85.1	87.4	87.8
Density 15 °C	ASTM D 4052-18	kg/m^3	733	735	737
Vapor pressure (DVPE)	ASTM D 5191-15	kPa	67.1	72.5	71.9
Vapor Lock Index	EN 228	–	903	949	951
Sulphur	ASTM D 4294-16e1	mg/kg	9	10	10
Lead	EN 237:2005	mg/L	<5.0	<5.0	<5.0
Existent Gums	ASTM D 381-12 (2017)	mg/100 mL	<0.1	<0.1	<0.1
Distillation					
Evaporated 70 °C (E70)		% *v/v*	33.1	32.0	33.2
Evaporated 100 °C (E100)		% *v/v*	56.3	55.9	57.2
Evaporated 150 °C (E150)	ASTM D 86-17	% *v/v*	82.7	82.6	83.8
Final Boiling Point		°C	194.8	191	198.0
Residue		% *v/v*	1.0	1.1	1.0
Hydrocarbons					
Olefins		% *v/v*	13.0	11.1	11.1
Aromatics		% *v/v*	27.9	24.4	24.4
Benzene		% *v/v*	0.5	0.7	0.7
Oxygen		% m/m		2.4	2.4
Methanol	EN ISO 22854:2016	% *v/v*	<0.1	<0.1	<0.1
Ethanol		% *v/v*	1.2	0.7	0.7
Isopropyl alcohol		% *v/v*	<0.5	<0.5	<0.5
ETBE		% *v/v*	7.48	13.5	13.5
Other oxygenated compounds		% *v/v*	<0.1	<0.1	<0.1

* American Society for Testing and Materials.

2.3. Test Protocol

Firstly, the aerodynamic and rolling resistance of the vehicle (road load coefficients) were calculated in order to be replicated in the chassis dyno. The determination of the road load coefficients was performed following the methodology proposed by Regulation 1151/2017 of the European Commission, sub-annex 4, paragraph 4 [29]. Tests were carried out on a straight and flat local road (CM-4117, Spain). The methodology consists in warming up the vehicle and subsequently accelerating up to 140 km/h; after that, coastdown is started with the gearbox in neutral until 20 km/h. Four repeats were performed in opposite directions and wind conditions and ambient temperature were measured. The coastdown time measurements were used to calculate the road load coefficients according to the mentioned regulation.

Coastdown coefficients were subsequently adapted to the chassis dyno in order to replicate real conditions on the road through the roller. For this task, the methodology followed the one proposed by Regulation 1151/2017 of the European commission, sub-annex 4, paragraph 7 and 8 [29].

All tests were carried out at 35 °C (ambient temperature set in the chassis dyno) and the total flow used in the dilution tunnel was 14 m^3/min. The cycle designed for these tests (Figure 2) comprises 10 full-load accelerations from 30 km/h to 162 km/h (toothsaw-type [6], all carried out in fourth gear), followed by two steady-state working points. Prior to the 10 toothsaw, a 10-min warming up at 80 km/h was carried out. Before each toothsaw acceleration, a 2-min period at 30 km/h is driven to ensure that all engine parameters are stable before starting the accelerations. Once the vehicle reaches 162 km/h, it remains at this velocity for 5 s, approximately. Finally, the vehicle slows down at constant deceleration to reach 30 km/h again.

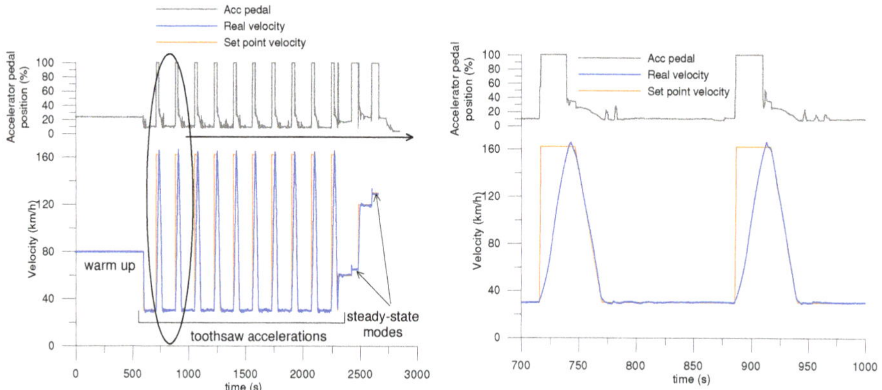

Figure 2. Driving cycle (**left**) and detail of the first accelerations (**right**).

The two steady-state operation modes were selected at 65 km/h and 129 km/h in fourth gear and full-load conditions (around 2500 rpm and 5000 rpm, respectively). The engine was maintained for one minute at each working mode and the results shown in this work are the average for this time period (after discarding the first 15 s, where the variables were not steady yet). Each mode was preceded by an intermediate mode at low load to ensure the vehicle is properly cooled down before carrying out the full-load modes. The whole cycle (setpoint and real velocity-time traces, along with the accelerator position) is shown in Figure 2 left. Moreover, in order to observe clearly the differences between targeted and real speed, Figure 2 right enlarges a small part of the cycle showing the first two toothsaw accelerations.

For the fuel changes, the fuel tank and fuel lines were drained and refilled with the new fuel. After this, a driving cycle (WLTC) was carried out to ensure a complete purge of the fuel supply system.

3. Results and Discussion

The most important performance parameters, such as acceleration, power output and fuel consumption, are firstly analysed for the three fuels at both engine conditions (toothsaw accelerations and steady-state modes). Boost pressure, spark timing and catalyst temperature are included in the discussion to explain the trends. Then, emissions are compared because of their strong dependence on some of the engine working parameters, which must be also considered to explain the effect of the fuel used. Regarding the accelerations, results presented in this section are the average of the 10 toothsaw accelerations. The error bars in the figures are defined as the confidence interval (95% confidence level).

3.1. Power Output and Acceleration

Maximizing power output and efficiency is the main goal of high-octane gasoline fuels, as detailed in the introduction. Figure 3 represents the acceleration and power output (power at the wheels) for the three gasolines tested during the toothsaw accelerations. Vehicle acceleration depends on the difference between the traction force in the wheels (which depends on the effective torque delivered by the engine) and the resistance forces, calculated through the coastdown test and imposed by the dyno. At each velocity, the traction force, and hence the vehicle acceleration, increases with the engine torque. The power delivered (Figure 3a) by the high-octane fuels (98 and 100 RON gasolines) is higher compared to 95 RON gasoline. This increase is not uniform throughout the whole velocity and octane ranges, but is sharper for the 100 RON fuel and when velocity is higher than 60–70 km/h (around 2500 rpm). This is consistent with the results of acceleration time (Figure 3b). As illustrated, the higher power delivered by the 100 RON gasoline produced the highest acceleration and thus the shortest time to reach the final velocity goal (162 km/h). 98 and 100 RON gasolines reduced the acceleration duration (compared to 95 RON) by 2.7% and 6.7%, respectively. During the steady-state modes, the power in the wheels (Figure 3) increased with the octane number of the fuels, and again the effect was more significant with the 100 RON gasoline. The 98 RON gasoline increased power by 3.4%, while this number increased up to 7.7% with the highest-octane gasoline (both compared to 95 RON gasoline).

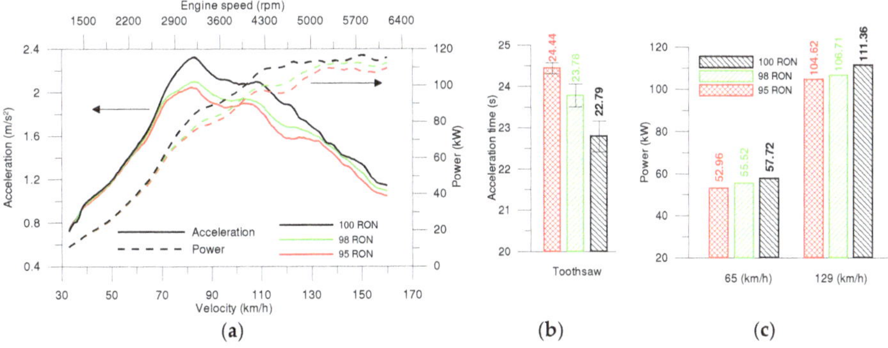

Figure 3. (**a**) Acceleration and power output along the toothsaw, (**b**) total acceleration time for toothsaw and (**c**) power output for steady-state modes.

As described in the introduction, current SI engines can adapt some operating parameters (boost pressure and spark timing advance, mainly) depending on the gasoline autoignition resistance in order to protect the engine against knocking and to reach the highest power output. Therefore, evaluating these parameters (and others that are affected by the operating parameters, such as the exhaust temperature) is necessary to explain the power and acceleration differences.

Turbocharging (increasing boost pressure) is one of the current tendencies in SI engines to improve the power output and efficiency. The intake air flow rate increases with the boost pressure; thus, more fuel can be admitted in the mixture, increasing the power. However, boost pressure is limited as it contributes to knocking due to the higher pressure and temperature at the beginning of the compression stroke. The tested vehicle incorporates a boost pressure control system (based on commanding a waste-gate valve in the turbine) managed by the ECU.

Figure 4 shows the boost pressure and the temperature upstream catalyst, which is directly related to the position of the waste-gate valve, for all the fuels and engine modes tested. Regarding the toothsaw accelerations (Figure 4a), boost pressure and air mass flow rate were higher for 100 RON gasoline, especially in the range between 70 to 130 km/h, with no significant differences between the other two fuels. This is because the higher knocking resistance of 100 RON gasoline allowed higher boosting without knocking. This velocity range (70–130 km/h) agrees well with the range where power and acceleration were the highest for the 100 RON fuel (Figure 3), which points to the boost pressure control system as the main factor contributing to the outstanding performance of this fuel.

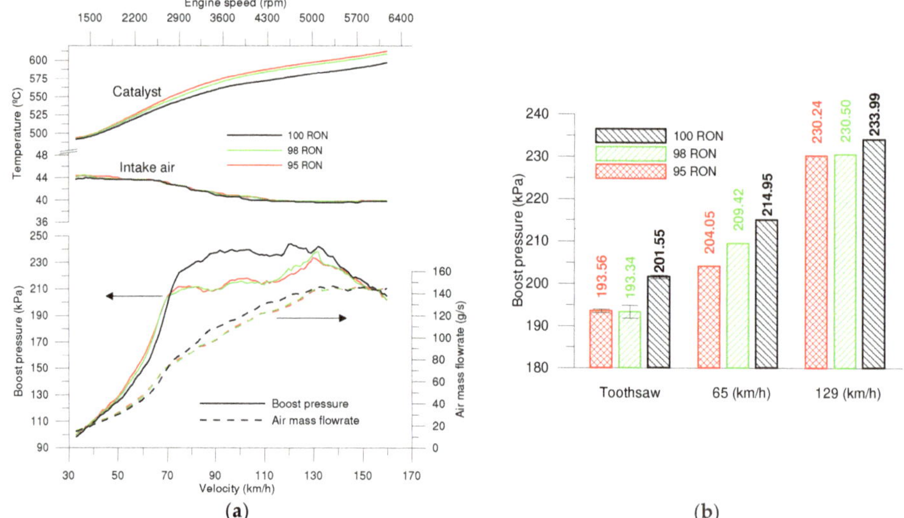

Figure 4. (**a**) absolute boost pressure, air mass flow rate, upstream catalyst temperature and intake air temperature along the toothsaw, (**b**) average absolute boost pressure along toothsaw and steady-state modes.

Boost pressure values were consistent with the exhaust temperature upstream of the three-way catalyst (Figure 4a). The higher boost pressure with the 100 RON gasoline is achieved by increasing the exhaust flow rate that passes through the turbine (i.e., decreasing the exhaust fraction that by-pass the turbine through the waste-gate valve). Since the gas temperature decreases in the turbine, as dictated by the gas dynamics in this device, the lower exhaust temperature with the 100 RON gasoline is a consequence of the lower waste-gate valve opening. As observed in Figure 4, the intake temperature was the same for all fuels and then could not contribute to the differences found in the exhaust gas temperature. Similar results were found at both steady-state modes (Figure 4b), with 100 RON gasoline having the highest boost pressure. In the 65 km/h mode, an increase in boost pressure with the 98 RON gasoline (compared to 95 RON) was observed as well, which indeed supports its higher power (Figure 3b).

Lambda sensor values are displayed in Figure 5 and reveal the calibration strategy of the engine; moreover, lambda values are decisive to explain the emissions. As it is typical in these sensors, there are two voltage levels (around 0.1 V and 0.9 V for lean and rich combustion, respectively) and a rapid, step-like transition zone around stoichiometric combustion. Along the toothsaw accelerations, the engine started running with oxygen excess (lean) at low engine speed (below 1500 rpm) for a short time. After pressing the gas pedal wide open, the throttle valve opens abruptly and there is a sudden increase in the air mass flow that lead to a brief lean operation. After that, the combustion moved to near-stoichiometric fuel-air ratio, and slightly richer as the speed increases up to 3300 rpm, approximately. From 3300 rpm onwards, the combustion became extremely rich. Consistently with the boost pressure trend, average lambda values scaled inversely with the octane number of the fuel (Figure 5b) indicating less rich combustion as the octane number is increased. This was demonstrated in the accelerations and in both steady-state modes.

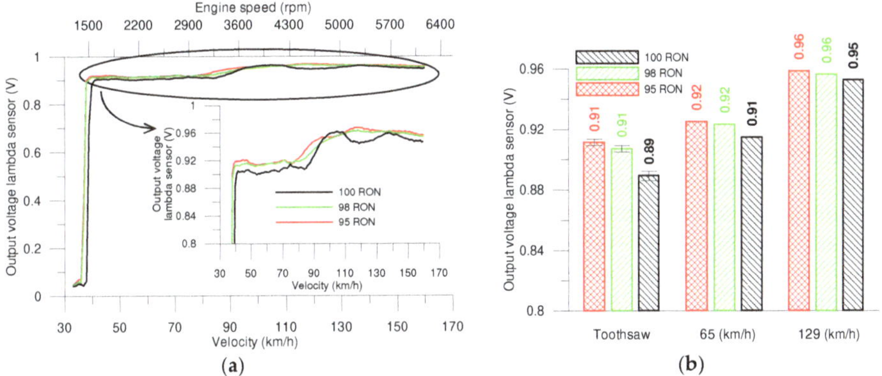

Figure 5. (**a**) Lambda values along the toothsaw, (**b**) average lambda values along toothsaw and steady-state modes.

Spark ignition advance is another parameter that affects the engine performance. As revised in the introduction section, there is an optimal ignition timing (MBT) that maximizes the output torque and the engine efficiency. SI engines do not usually work at MBT conditions, at least not in the whole speed and torque range, since this would lead to severe permanent knocking. Hence, spark timing advance is typically delayed with respect to MBT and this reduces the effective torque available. In the toothsaw accelerations (Figure 6a), two different trends were observed depending on the speed range. First, from the beginning to 70 km/h the spark ignition is more advanced as the octane number of the fuel is increased. Then, from 70 km/h onwards, the differences in the spark timing of the fuels are less significant, and the more advanced spark ignition remains only with the 98 RON gasoline. As observed, the cut point velocity of both ranges (70 km/h) coincides well with the velocity from which the boost pressure (Figure 4a) starts to be sensitive to the fuel. This could indicate that the internal control algorithms of both variables under transient accelerations are coupled in this engine. By contrast, when the steady-state modes were tested (Figure 6b), a slightly more advanced spark timing was measured with the increase of the octane number (around 1 °CA for the 100 RON gasoline, compared to 95 RON, with a linear trend between fuels).

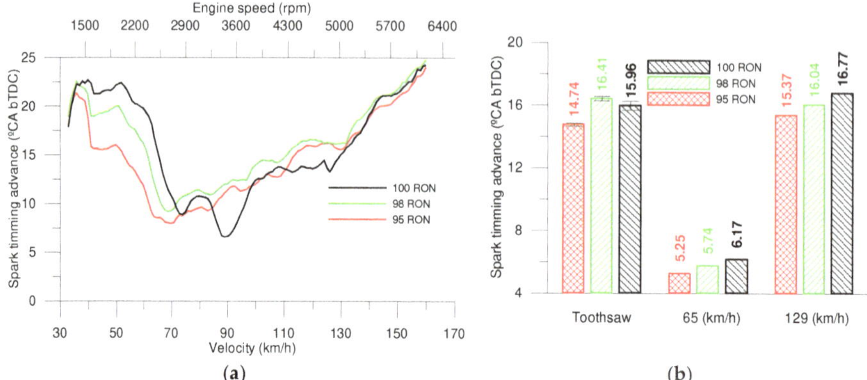

Figure 6. (**a**) Spark ignition advance along the toothsaw, (**b**) average spark ignition advance along toothsaw and steady-state modes.

3.2. Specific Fuel Consumption

The specific fuel consumption is shown in Figure 7. Fuel consumption is calculated based on a carbon balance method, as stated in the section devoted to the experimental setup. Since the three tested fuels have close heating values, the specific fuel consumption is inversely proportional to the engine efficiency. There is a marked decrease in the specific fuel consumption with the increase in octane number, this effect being more notable in the toothsaw accelerations. Compared to 95 RON gasoline, 98 and 100 RON gasolines decreased the specific fuel consumption in the toothsaw accelerations by 5% and 12%, respectively; these figures were reduced to 2.8% and 8.3% (average) in the steady-state modes. As observed, the fuel save of the 100 RON gasoline was superior than that of the 98 RON gasoline, even in relative terms. By contrast, other authors [5] reported a linear benefit of increasing RON on the fuel consumption when running driving cycles. This is because knocking is a phenomenon extremely dependent on the engine design, engine calibration and operating conditions. Although higher RON may be positive for a wide range of vehicles equipped with knock control devices, the exact magnitude and the proportionality of the benefits depends on the combination of these factors.

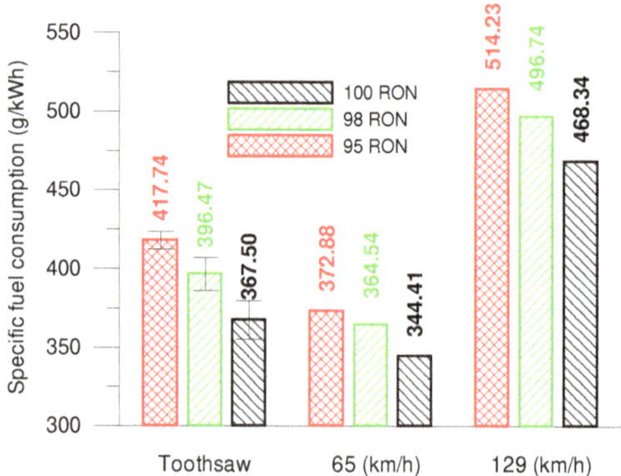

Figure 7. Average specific fuel consumption along toothsaw and steady-state modes.

The positive effect of the octane number on the fuel economy and the efficiency is a combination of two factors. First, under rich conditions (Figure 5), the higher intake air flow rate with high-octane fuels (Figure 4a), derived from the higher boost pressure, leads to more energy release from the fuel which increases the indicated torque and consequently the mechanical efficiency. Second, the more advanced ignition (Figure 6) increases the thermal efficiency of the cycle.

3.3. Gaseous Emissions

During the whole cycle (toothsaw and steady-state modes), the gaseous emissions were measured continuously. Figure 8 shows CO_2, CO, THC and NOx emissions during the toothsaw acceleration. Several regions are observed in the figure. From the beginning to 55–60 km/h, there were no significant CO and THC emissions. This is a consequence of the initial lean operation (as indicated by the lambda sensor, Figure 5a) and the subsequent operation at near-stoichiometric conditions (optimal conditions for the three-way catalyst), where the CO and THC generated in the combustion chamber were abated in the three-way catalyst. In this first region, CO_2 emission increased slowly, proportionally to the power. Most of the total NOx in the accelerations were emitted in this region, where NOx reached a peak (see Figure 7b) because the catalyst could not reduce the engine-out NOx under the initial lean conditions. In the second region (from 55 km/h to around 90 km/h), both CO_2 and CO increased with the velocity, being this increase approximately linear and steeper in the case of CO_2. THC emissions also increased. These trends agree well with the lambda sensor values in this velocity range (55–90 km/h), which indicated a progressive and slow transition from near-stoichiometric to slightly rich combustion. There is a last region (from 90 km/h) where the CO_2 emission stabilized, while THC and CO continued to increase with the velocity. The rate of increase was higher than in the previous region (this is more evident in the CO trace, Figure 8a, which reached even higher values than CO_2) because more fuel is consumed to respond to the acceleration demand, but it is not oxidized completely to CO_2. Again, this is consistent with the lambda values, which indicated a faster shift towards much richer combustion. Even the final stabilization of the CO emission agrees with the stabilization of the lambda values.

For 100 RON gasoline, in the last velocity region there was a trade-off between CO_2 and CO/THC (when the former increased the others decreased, and vice versa), especially evidenced by fluctuations around a trend line in the emission flow rate profiles (Figure 8a). Under oxygen-limited conditions, fuel molecules compete for the available oxygen in the combustion chamber: when oxygen becomes more accessibility to fuel molecules (more homogeneous mixture, higher turbulence, etc.), CO_2 increases and CO/THC decrease. Also, the three-way catalyst plays a role in the final CO/CO_2 values, since in the absence of oxygen (rich combustion) a water gas shift reaction may occur [30]. These processes in the combustion chamber and the catalyst may be the cause for the aforementioned trends for CO and CO_2.

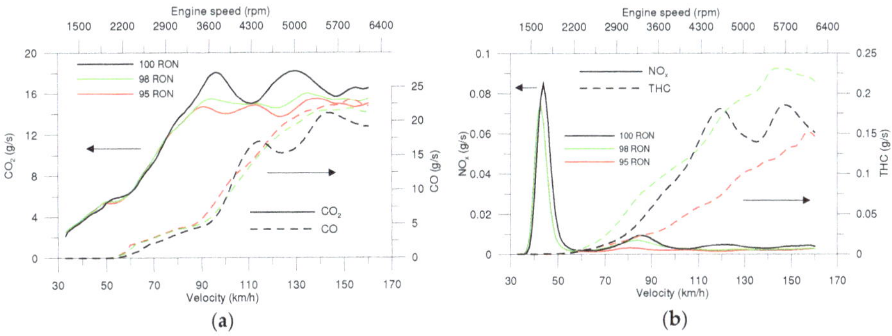

Figure 8. CO_2 and CO emissions (**a**) and NO_x and total hydrocarbon (THC) emissions (**b**) along toothsaw.

Regarding the effect of the fuel, the CO_2 flow rate increased with the octane number, whereas CO decreased (Figure 8a) at velocities higher than 70 km/h (approximately). The main responsible factors are the higher intake air flow rate in the case of 100 RON gasoline, compared to 95 RON (leading to lower fuel enrichment), and the more advanced ignited in the case of 98 RON gasoline, both factors contributing towards a more complete combustion. When the specific CO_2 and CO emissions (g/kWh) are discussed (Figure 9a,b), CO_2 was not affected by the fuel because the octane number contributed to higher power and this counteracted the higher CO_2 flow rate, but CO was greatly reduced with increasing octane number. Compared to 95 RON gasoline, 98 and 100 RON gasolines decreased the specific CO emission by 9% and 20%, respectively, and the trend is consolidated in both the accelerations and the steady-state modes.

There was not a clear tendency of specific THC emissions (Figure 9d) with the octane number of the fuel, with the intermediate RON fuel showing the highest values in all the tests. Nevertheless, the fuel effect, if any, is probably masked by the very low concentration emitted compared to that of CO. Low THC and high CO emissions confirm that the oxygen available on the combustion chamber is enough to initiate the combustion reactions of the fuel hydrocarbons but not to complete them. Finally, NOx emissions were in general higher for higher octane fuels (except in the steady-state mode at 65 km/h, Figure 9c), probably on account of the higher combustion temperatures derived from the lower fuel enrichment and the higher intake temperature (due to the increased boost pressure).

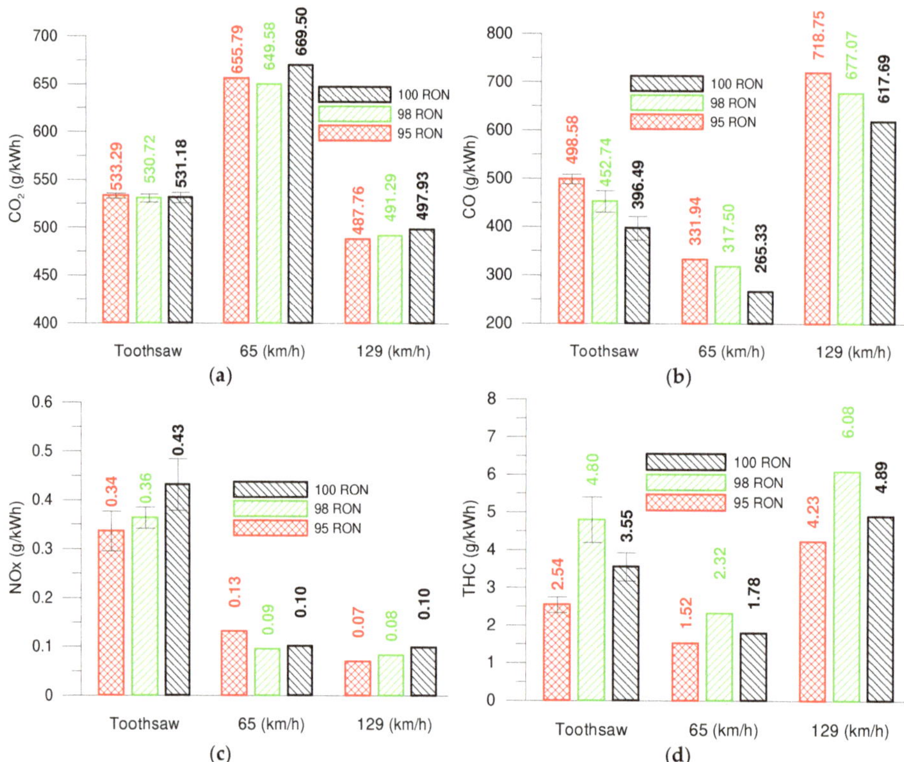

Figure 9. CO_2 (**a**), CO (**b**), NOx (**c**) and THC (**d**) emissions.

4. Conclusions

The potential of high-octane gasolines to increase fuel economy and thermal efficiency has been evaluated in a Euro 6b passenger car running at full load accelerations and two steady conditions. The results have proved that high octane number can be exploited by current and future SI engine technologies, with no penalization or some benefits in gaseous emissions.

High-octane fuels led to increased power and acceleration and reduced specific fuel consumption. The effect of the octane number was nonlinear, with 98 RON improving the base case (95 RON) moderately, but 100 RON improving it greatly. As an example, 98 and 100 RON gasolines reduced the acceleration duration (compared to 95 RON) by 2.7% and 6.7%, respectively. Moreover, the power gain was not uniform in the whole speed range: at low velocity there was no significant effect whereas in the middle range the effect was highly marked. The main reasons supporting these results were the higher boost pressure (specially with 100 RON gasoline) and the advanced spark timing (specially with 98 RON). The electronic control unit of the engine manages both parameters conveniently to avoid knocking, which is detected through a knock sensor equipped in the vehicle.

Octane number did not affect THC or NOx emissions significantly, although the latter slightly increased with the 100 RON fuel. Regardless of the fuel, a NOx peak was manifested at the start of the accelerations because of a momentary lean operation. CO emission was reduced with the octane number, which was a consequence of the higher air-to-fuel ratio derived from the higher boost pressure.

Author Contributions: Conceptualization, D.C. and J.D.; Data curation, Á.R.; Formal analysis, J.R.-F. and Á.R.; Funding acquisition, J.D.; Investigation, Á.R. and J.B.; Methodology, J.R.-F., Á.R., J.B. and D.C.; Project administration, J.R.-F. and J.D.; Resources, J.D.; Software, Á.R.; Supervision, D.C. and J.D.; Validation, Á.R.; Visualization, D.C. and J.D.; Writing—original draft, J.R.-F., Á.R. and J.B.; Writing—review & editing, J.R.-F., Á.R., J.B., D.C. and J.D. All authors have read and agreed to the published version of the manuscript.

Funding: This research received no external funding.

Conflicts of Interest: The authors declare no conflict of interest.

References

1. European Commission. A European Green Deal: Striving to be the First Climate-Neutral Continent. 2019. Available online: https://ec.europa.eu/info/strategy/priorities-2019-2024/european-green-deal_en (accessed on 6 July 2020).
2. European Environment Agency. Greenhouse Gas Emissions from Transport in Europe. 2019. Available online: https://www.eea.europa.eu/data-and-maps/indicators/transport-emissions-of-greenhouse-gases/transport-emissions-of-greenhouse-gases-12 (accessed on 6 July 2020).
3. Knuutila, L. Blending Strategies and Process Modification for the Future Gasoline Production. Master's Thesis, Programme in Advance Energy Solutions. Aalto University, Espoo, Finland, 2019.
4. Kalghatgi, G. Developments in internal combustion engines and implications for combustion science and future transport fuels. *Proc. Combust. Inst.* **2015**, *35*, 101–115. [CrossRef]
5. Williams, J.; Hamje, H. *Testing and Modelling the Effect of High Octane Petrols on an Adapted Vehicle*; Technical Report No. 8/20; Concawe: Brussels, Belgium, 2020. Available online: https://www.concawe.eu/wp-content/uploads/Rpt_20-8.pdf (accessed on 6 July 2020).
6. Stradling, R.; Williams, J.; Hamje, H.; Rickeard, D. Effect of Octane on Performance, Energy Consumption and Emissions of Two Euro 4 Passenger Cars. *Transp. Res. Procedia* **2016**, *14*, 3159–3168. [CrossRef]
7. Kalghatgi, G. Knock onset, knock intensity, superknock and preignition in spark ignition engines. *Int. J. Engine Res.* **2017**, *19*, 7–20. [CrossRef]
8. Pan, J.; Shu, G.; Wei, H. Interaction of Flame Propagation and Pressure Waves during Knocking Combustion in Spark-Ignition Engines. *Combust. Sci. Technol.* **2014**, *186*, 192–209. [CrossRef]
9. Kalghatgi, G.T.; Bradley, D. Pre-ignition and 'super-knock' in turbo-charged spark-ignition engines. *Int. J. Engine Res.* **2012**, *13*, 399–414. [CrossRef]
10. Hamilton, L.J.; Rostedt, M.G.; Caton, P.A.; Cowart, J.S. Pre-Ignition Characteristics of Ethanol and E85 in a Spark Ignition Engine. *SAE Int. J. Fuels Lubr.* **2008**, *1*, 145–154. [CrossRef]
11. Dahnz, C.; Han, K.-M.; Spicher, U.; Magar, M.; Schiessl, R.; Maas, U. Investigations on Pre-Ignition in Highly Supercharged SI Engines. *SAE Int. J. Engines* **2010**, *3*, 214–224. [CrossRef]
12. Manz, P.W.; Daniel, M.; Jippa, K.N. Pre-ignition in highly charged turbo-charged engines. Analysis procedure and results. In Proceedings of the 8th International Symposium on Internal Combustion Diagnostics, Baden-Baden, Germany, 10–11 June 2008.
13. Wang, Z.; Liu, H.; Song, T.; Qi, Y.; He, X.; Shuai, S.; Wang, J. Relationship between super-knock and pre-ignition. *Int. J. Engine Res.* **2014**, *16*, 166–180. [CrossRef]

14. Ceschini, L.; Morri, A.; Balducci, E.; Cavina, N.; Rojo, N.; Calogero, L.; Poggio, L. Experimental observations of engine piston damage induced by knocking combustion. *Mater. Des.* **2017**, *114*, 312–325. [CrossRef]
15. Bi, F.; Ma, T.; Wang, X. Development of a novel knock characteristic detection method for gasoline engines based on wavelet-denoising and EMD decomposition. *Mech. Syst. Signal Process.* **2019**, *117*, 517–536. [CrossRef]
16. Feng, D.; Buresheid, K.; Zhao, H.; Wei, H.; Chen, C. Investigation of lubricant induced pre-ignition and knocking combustion in an optical spark ignition engine. *Proc. Combust. Inst.* **2019**, *37*, 4901–4910. [CrossRef]
17. Shao, J.; Rutland, C.J. Modeling Investigation of Different Methods to Suppress Engine Knock on a Small Spark Ignition Engine. *J. Eng. Gas Turbines Power* **2015**, *137*, 061506. [CrossRef]
18. Lim, G.; Lee, S.; Park, C.; Choi, Y.; Kim, C. Effect of ignition timing retard strategy on NOx reduction in hydrogen-compressed natural gas blend engine with increased compression ratio. *Int. J. Hydrogen Energy* **2014**, *39*, 2399–2408. [CrossRef]
19. Szybist, J.; Foster, M.; Moore, W.R.; Confer, K.; Youngquist, A.; Wagner, R.M. Investigation of Knock Limited Compression Ratio of Ethanol Gasoline Blends. *SAE Paper* **2010**. [CrossRef]
20. Grandin, B.; Denbratt, I.; Bood, J.; Brackmann, C.; Bengtsson, P.-E.; Gogan, A.; Mauss, F.; Sundén, B. Heat Release in the End-Gas Prior to Knock in Lean, Rich and Stoichiometric Mixtures with and without EGR. *SAE Paper* **2002**. [CrossRef]
21. Wei, H.; Zhu, T.; Shu, G.; Tan, L.; Wang, Y. Gasoline engine exhaust gas recirculation—A review. *Appl. Energy* **2012**, *99*, 534–544. [CrossRef]
22. Grandin, B.; Ångström, H.-E. Replacing Fuel Enrichment in a Turbo Charged SI Engine: Lean Burn or Cooled EGR. *SAE Paper* **1999**. [CrossRef]
23. Wang, Z.; Liu, H.; Reitz, R.D. Knocking combustion in spark-ignition engines. *Prog. Energy Combust. Sci.* **2017**, *61*, 78–112. [CrossRef]
24. Zhen, X.; Wang, Y.; Xu, S.; Zhu, Y.; Tao, C.; Xu, T.; Song, M. The engine knock analysis—An overview. *Appl. Energy* **2012**, *92*, 628–636. [CrossRef]
25. Li, T.; Yin, T.; Wang, B.; Qiao, X. Determination of Knock Limited Spark Advance in Engine Cycle Simulation. In Proceedings of the International Symposium on Diagnostics and Modeling of Combustion in Internal Combustion Engines, Okayama, Japan, 25–28 July 2017; Japan Society of Mechanical Engineers: Tokyo, Japan, 2017; p. A103.
26. Shen, Y.T.; Wang, J.Z.; Shuai, S.J.; Wang, J.X. Effects of octane number on gasoline engine performance. *Chin. Int. Combust. Engine Eng.* **2008**, *29*, 52–56.
27. Shuai, S.-J.; Wang, Y.; Li, X.; Fu, H.; Xiao, J. Impact of Octane Number on Fuel Efficiency of Modern Vehicles. *SAE Int. J. Fuels Lubr.* **2013**, *6*, 702–712. [CrossRef]
28. European Automobile Manufacturers Association (ACEA); European Medicines Agency; JAMA. *Worldwide Fuel Charter: Gasoline and Diesel Fuel*, 6th ed. 2019. Available online: https://www.acea.be/uploads/publications/WWFC_19_gasoline_diesel.pdf (accessed on 6 July 2020).
29. European Commission. Regulation 2017/1151 of 1 June 2017 supplementing Regulation (EC) No 715/2007 (...) to emissions from light passenger and commercial vehicles (Euro 5 and Euro 6) (...). *Off. J. Eur. Union* 02017R1151, 1-839. **2017**.
30. Han, Z.; Wang, J.; Yan, H.; Shen, M.; Wang, J.; Wang, W.; Yang, M. Performance of dynamic oxygen storage capacity, water–gas shift and steam reforming reactions over Pd-only three-way catalysts. *Catal. Today* **2010**, *158*, 481–489. [CrossRef]

Review

Alternative Fuels for Internal Combustion Engines

Jorge Martins * and **F. P. Brito**

MEtRICs, Universidade do Minho, 4710-057 Braga, Portugal; francisco@dem.uminho.pt
* Correspondence: jmartins@dem.uminho.pt

Received: 20 May 2020; Accepted: 31 July 2020; Published: 6 August 2020

Abstract: The recent transport electrification trend is pushing governments to limit the future use of Internal Combustion Engines (ICEs). However, the rationale for this strong limitation is frequently not sufficiently addressed or justified. The problem does not seem to lie within the engines nor with the combustion by themselves but seemingly, rather with the rise in greenhouse gases (GHG), namely CO_2, rejected to the atmosphere. However, it is frequent that the distinction between fossil CO_2 and renewable CO_2 production is not made, or even between CO_2 emissions and pollutant emissions. The present revision paper discusses and introduces different alternative fuels that can be burned in IC Engines and would eliminate, or substantially reduce the emission of fossil CO_2 into the atmosphere. These may be non-carbon fuels such as hydrogen or ammonia, or biofuels such as alcohols, ethers or esters, including synthetic fuels. There are also other types of fuels that may be used, such as those based on turpentine or even glycerin which could maintain ICEs as a valuable option for transportation.

Keywords: biofuels; fuels; synthetic fuels; internal combustion engine; alternative fuels

---

1. Introduction

Since the beginning of the industrial age, the burning of fossil fuels has been releasing to the atmosphere the carbon that was slowly sequestered more than 50 million years ago and stored as coal, oil, natural gas and other types of fossil fuels sources such as shale gas and shale oil. The combustion of these fuels produces, besides pollutants, carbon dioxide (CO_2) which has a major effect of trapping the solar heat within the atmosphere (greenhouse effect), is expected to aim to the warming up of the Earth and the severity of the climate [1].

The general public and most policy makers perceive electric vehicles as a good alternative to fossil fuel-based transportation [1–3]. However, regarding emissions linked to vehicle use, electric vehicles are as green as the electricity they consume [4]. In a country such as Poland [5] or Australia, where most of the electricity is produced from coal, the running of electric vehicles will increase the contribution to the greenhouse effect comparing to the same car running in Norway or Brazil, where most of the electricity is produced from renewable energy sources [6]. Additionally, the evaluation of the sustainability of EVs must take into account the whole life cycle of the vehicle, including sensible issues such as the mining of the materials used in batteries and electric motors, as well as battery end-of-life [1,4,7,8]. Nonetheless, although it is a hot topic with a lot of promising developments [1], the detailed sustainability comparison between conventional and electric mobility is out of the scope of the present study and is only treated to provide some background.

Now, a large number of countries, regions and cities are proposing the ban on so-called "conventional" vehicles within the next decades [9,10] Normally, what policy makers refer to when using this term are cars, buses or lorries propelled by an Internal Combustion Engine (ICE) that burns a fossil fuel. So, if one of these specifications is eliminated from a vehicle, it will no longer be "conventional". This leaves three types of vehicles: electric, hybrid-electric and alternative-fuel burning.

Nevertheless, some references to the future ruling out of so-called "polluting" vehicles are specified in terms of a ban on Internal Combustion (IC) Engines altogether, or in terms of a ban on vehicles that burn specific fuels such as diesel or gasoline [4]. This is surprising, as it is accompanied by a growing acceptance of Plug-in Hybrid Electric Vehicles (PHEVs) that are able to run for several tens of kilometres without burning fossil fuel and therefore without producing pollution locally (in city centres, for example) [11]. But these vehicles rely on ICEs for some of their operation, namely in long trips, once the battery has been depleted. Under these conditions, the ICE usually runs on fossil fuel, therefore producing a non-negligible amount of pollutants and CO_2 during operation [12].

Therefore, what policy-makers likely aim with the strong limitations to conventional vehicles is the reduction of the emission of fossil CO_2 and the elimination of pollutant emissions within the city limits, which can be jointly achieved by the use of electrified (PHEV) vehicles [3] This seems to be the direction most OEMs (original equipment manufacturers) are taking. Volvo, for example, had vowed to stop developing "conventional" (non-electrified) vehicles from 2019, only hybrid and battery electric ones [13]. As of 2020 this has been mostly fulfilled although some of them are only "mild-hybrids" which have a bigger starter generator that displays some braking energy recovery and limited engine assist [14]. Similar commitments can be seen in other OEMs. Nevertheless, the announcements and proposed timeframes towards electrification or even ICE development abandonment are not always completely fulfilled [15,16].

However, GHG (greenhouse gases) reduction in the transport sector can hugely benefit from the use of ICE using CO_2 neutral fuels [17]. This is particularly important in sectors, such as heavy-duty and aviation, where energy density plays an important role. A recent paper about future trends in transport [4] calculated that, with the present battery technology, electric passenger airplanes would require between 14 and 31 times its maximum take-off weight in batteries to store the energy that they usually carry as jet fuel. Also, the time for battery charging, using 80 Tesla superchargers would take over one day to fully recharge the battery equivalent of an Airbus 320 fuel tank. In terms of large vessels, the 170 GWh of energy that some these types of container ships carry in their tanks to power the engines would require batteries over five times their dead weight and would take years to recharge [4]. These estimations seem to be too pessimistic as they do not take into account the difference in efficiencies between electric motors and IC engines, which would cut in half the energy needed, but are high enough to illustrate the impracticality of electrification for large carriers to travel over very long distances unless highly disruptive changes take place in battery technology. But renewable fuels and/or biofuels may be good propositions for these types of transport.

2. Hybrid Vehicles

Electrified vehicles need not specifically be battery electric vehicles (BEV), but there are various levels of hybrid vehicles, from the standard plug-in (parallel) hybrids, in which the engine provides mechanical traction power, to the extended-range (series) hybrids, in which traction is solely made by electric motors and the engine merely generates electricity, to fuel-cell hybrids. All these types of hybrid cars use some type of fuel that is burned in an ICE, with the exception of the fuel-cell hybrids. These latter vehicles use hydrogen (or another hydrogen-rich fuel such as alcohols or ammonia) which does not burn, but goes through a different process normally involving a catalysis through a Proton Exchange Membrane (PEM) producing electricity, water or water and CO_2, when the molecule of the fuel also comprises carbon atoms [18,19]. While hybrid systems tend to duplicate systems, which might be a disadvantage in terms of cost and maintenance needs, it might provide a positive trade-off in the short to midterm, as it allows to minimize the main current limitations of electric mobility: energy storage cost, density, reliability and charging time [8,20]. While some of these limitations are no longer critical for small urban vehicles, which do not need large storage, they are still critical for driving patterns needing frequent long trips, which would require huge, expensive energy storage systems needing long recharging times or very high power fast charging stations, which also have their own challenges [21]. For instance, the authors reported that the Plug-in hybrid well-to-wheel

CO_2 emissions may become negligible for cases where the long trips are less than 25% of the total mileage [12]. This was reported for the case where a compact and efficient range extender with two different operating conditions (one for efficiency another one for extra power) was implemented. Moreover, this configuration would allow low fuel consumption, fairly low complexity (comparatively to parallel hybrids) and low system cost [12].

Of course, the real-world emissions of vehicles are different from the reported emissions of new vehicles. Real driving emissions tests tried to address this issue and are now part of the emissions certification process. They are done under charge depleting (CD—mostly electric mode) and charge sustaining (CS—driving with a stable low state of charge using mostly the engine) modes. The official fuel consumption and emissions is a weighted average between the CD and CS modes, with the weighting factor being the so-called "utility factor" (UF) [22]. The UF used in Europe is based on the driving statistics described by the SAE J2841 standard [23]. But these tests are made to new model cars, not to cars being currently driven in roads. There is a loophole in this certification because some users will not use plug-in capability of the vehicle as often as desirable, relying excessively on the charge sustaining mode. Of course, to attain a more realistic measure of the emissions of fleet vehicles, the data mining of this information would be required in these vehicles, but this is still not the case. However, as long as there is an economic incentive in terms of energy cost in order to use electricity, it seems that a hybrid architecture would be well suited for a gradual transition towards full BEV mobility once their challenges have been overcome. Now, the sustainability of hybrid vehicles could be further improved by the use of fuels with lower GHG and pollutant emissions, as discussed in continuation.

3. Fuels

Fuels for IC engines (and fuel cells) are usually a combination of hydrogen (H) and carbon (C) atoms, but occasionally the fuel may also have other elements such as oxygen (O) or nitrogen (N). Fossil fuels (such as petrol or diesel) are, usually, a mixture of different components (hydrocarbons) composed of H and C. Each component has its own physical properties, such as density, boiling temperature and heating value (HV). One of the problems of IC engines is the very high standard for exhaust emissions which requires expensive and bulky after-treatment of the exhaust gases, which also reduces the fuel efficiency of the vehicle.

If a fuel contains oxygen or nitrogen, as these elements do not burn, its HV is lower than others composed just by carbon and hydrogen. In fact, when an alcohol (composed of C, H and O) burns, the oxygen atoms present in the burned gases are in the form of CO_2 or H_2O, mostly the latter [23,24].

Liquid fuels are more appropriate for vehicle propulsion. They have a very high energy density in terms of mass and volume, enabling the vehicles to have an enormous range. Gaseous fuels require pressurized tanks and have a much lower energy density, resulting in much larger and heavier tanks for the same amount of stored energy (Figure 1, [25]).

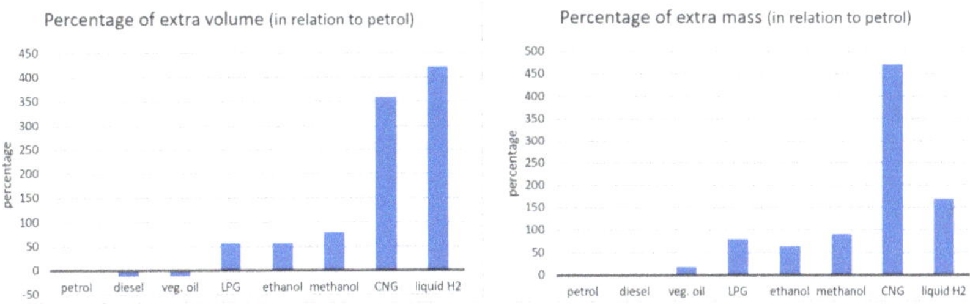

Figure 1. Percent increase of volume and mass of fuel tanks in relation to petrol (55 L tank), [25].

Battery electric vehicles, for example, require huge volumes for their batteries (Table 1), adding mass and cost [8]. Nevertheless, the advances in battery technology have been slow but steady, with emerging technologies gaining traction. Examples are the use of high capacity cathode (e.g., metal oxide) and anode materials, electrolytes with high oxidation potential and metal-air batteries which replace the positive electrode with an air electrode [26]. Another advantage of liquid fuels is their straightforward and fast refuelling. Gaseous fuels are more complicated and take longer to refuel, while electric vehicles require complex, expensive and time-consuming procedures.

Table 1. Storage energy in volume for fuels and batteries [25].

Fuel	Stored Energy (MJ/L)
diesel	36
petrol	33
biodiesel	33
LPG	25
LNG (@ −162 °C)	22
ethanol	21
methanol	16
CNG (@300 bar)	12
H_2 (liq. @ −253 °C)	8.5
H_2 (comp. @ 250 bar)	2.5
battery Li-ion	0.9–1.35
battery Pb-acid	0.3

In general, in terms of CO_2 production of fossil fuels, the more carbon the molecule has, the higher is the production of this gas. Table 2 shows the potential for CO_2 production of various fuels (in terms of LHV) compared to petrol (100). Please note that hydrogen is not a natural fuel, so it has to be produced from other sources, that may be fossil, therefore generating CO_2, not in its burning, but during its production.

Table 2. Potential for CO_2 production of various fuels (adapted from [27]).

Fuel	CO_2 Emissions
Petrol	100
Diesel	102
LPG	87
Natural gas	75
Hydrogen	0

4. Biofuels

It is important to know whether a fuel is based on renewable energy, such as crops, as when it burns it does not increase the level of fossil CO_2 in the atmosphere but marginally (if process and process emissions are taken into account [28], so there is an important division between different fuels is if they are generated from fossil or from renewable sources (biofuels) (Figure 2).

However, as it will be seen later, sometimes it is difficult to assess whether the fuel is "bio" (from renewable energy) or not. For example, biodiesel is seen as a biofuel, but the 10% of methanol required for the transesterification process is usually produced from natural gas, a fossil fuel (Figure 2). Some fuels (such as hydrogen or ammonia) can be produced from fossil sources (hydrogen from natural gas or oil) but they can also be produced entirely from renewable sources. This the case for the hydrogen produced from hydrolysis of water using renewable electricity such as solar or wind, as well as photochemical cells. Other so-called solar fuels can also incorporate CO_2 using the same methods [29]. Also, biomass can be converted to other more convenient fuel forms through the incorporation of solar energy, which aids the pyrolysis process [30].

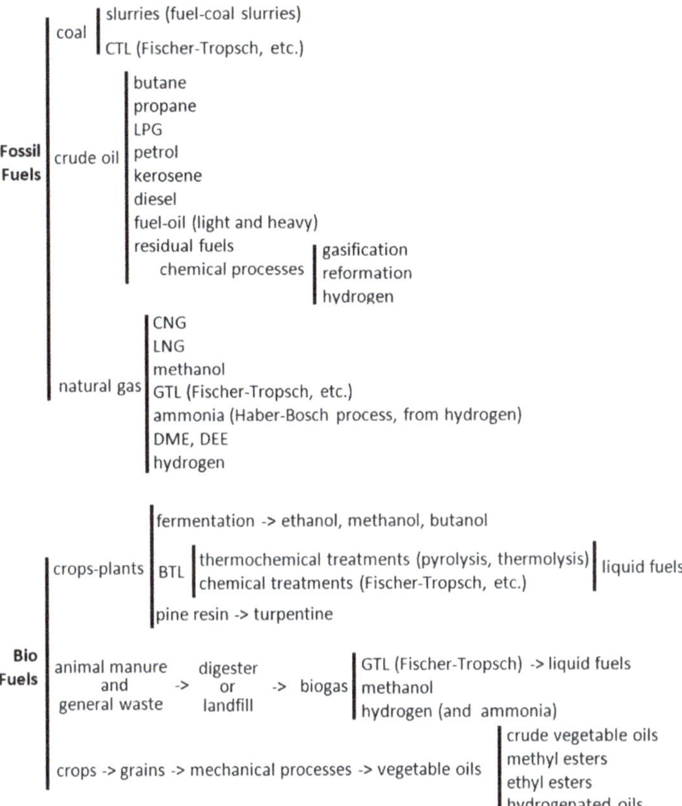

Figure 2. Types of fossil and biofuels that can be used in Internal Combustion Engines [25].

This brings another discussion about the fuels, because some of them (such as hydrogen) can be seen as "energy carriers" rather than "energy sources". Electricity is an energy carrier, as it may be produced from different energies sources (renewables such as solar and wind or fossil such as carbon or nuclear) in a specific location, but then it is transported to the place where the energy is needed, such as houses. The electricity may be transported over large distances by the electrical network or it may be transported by the vehicles within chemical-electric batteries. In this respect, hydrogen, ammonia and other synthetic fuels (such as Fischer-Tropsch petrol or diesel) can also be considered "energy carriers". If these fuels are produced entirely from renewable sources, then they are considered biofuels. So, these referred fuels (hydrogen, ammonia, synthetic hydrocarbons) may be considered fossil fuels, if they are produced from fossil origins, or they can be partially or entirely considered biofuels. Therefore, the SAME FUEL can be considered a fossil fuel or a biofuel. The methane present in the natural gas is fossil, whereas the same methane contained in the biogas is considered biofuel.

Important Properties

A list of the most common properties for different fuels gathered from several sources [27–38] can be seen in Table 3. Fuel density and the heating value are important for the determination of the quantity of energy available in the fuel, in terms of volume or mass. The heating value can be specified in terms of higher heating value (HHV) or lower heating value (LHV). The difference between HHV and LHV is the heat related to the condensation of the produced water. Obviously, for carbon, HHV has the same value than LHV, as there is no water produced by its combustion.

Table 3. Properties for some fuels (adapted from [27–38]).

Fuel	Chemical Formula	ρ_{Liquid} (kg/m^3)	$T_{Boiling}$ (°C)	Latent Heat of Vaporization (kJ/kg)	$T_{ignition}$ (°C)	$T_{adiabatic}$ (°C)	LHV (MJ/kg)	LHV_{Liquid} (MJ/L)	$HHV_{Mixture}$ (kJ/L)	A/F_{stoich}	RON	CN	Viscosity (@40 °C) (cSt)	Flash Point (°C)	Reid Vapour Pressure (@38 °C) (kPa)	Flammability Limits (% vol.)	O_2 (%)
Acetylene	C_2H_2	621	−84	614	305	2334	48.5		4.0	13.2	40					2.5–75	
Ammonia	NH_3	682	−33	1370	650	1803	18.6		2.8	6.1	110		0.3	-		15–28	
Av gas		715	25–170		440		44	32		~14.5	115			−40	30–90	1.5–7.6	
Biodiesel	$C_{17}H_{32}O_2$	850–885	250–350		220	2000	37	33		~13	-	45–65	3.5–5.5	62			~11
Biogas	CH_4	-	−162	510	580	1954	24		3.1	17	120						
Butane	C_4H_{10}	580	−0.6	386	405	1975	46.5	26	4.2	15.6	102					1.8–8.4	
Carbon	C				700		33			11.5							
Carbon monoxide	CO				609	2120		12		2.4						12.5–74	
Coal				-	450		27			11.5	-						
DEE	C_2H_5 O C_2H_5	714	34		160	1980	34	34		11.1		>125	0.23		110	1.9–36	
Diesel	$C_nH_{1.8n}$	820–870	180–360	~270	~210		43	36	3.9	~14	-	45–55	2.0–3.5	60–80	<1.5	0.6–8	
DMC	$C_3H_6O_3$	1079	90	418	458		15.8	17		4.64	109			16.7	10.8	4.2–12.9	53.3
DME	CH_3 O CH_3	667	−25	375	320	2020	28.8	19	3.4	9.0	-	55–60	0.18	−41	800	3.4–19	35
DMF	CH_3 C_4H_2 O CH_3	895	93	340	286		33.7		3.4	10.7	119			67		3.4–18.6	
DMM	CH_3**O** CH_2**O** CH_3	865		319	237		22.4					30					42.1
ETBE	$C_6H_{14}O$	770			307					12.2				−25		1.4–10	
Ethanol	C_2H_5 OH	790	78	900	365	1965	26.8	21		9.0	110	8	1.5	12	16	4.3–19	35
FAGE		962					34.2			11.2		70	11.9				
Fuel oil ("thick")		~960	>180	~230	~260	1995	41	36		~14	-					0.7–5	
F-T diesel	C_nH_{2n+2}	785	175–355		315		43.9			15.0		79	3.5				
Glycerine	$C_3H_5(OH)_3$	1260	290	670	390		19			4.2		5		170		3–19	52.2
HVO		776					43.9			15.0		82	2.65				
Hydrogen	H_2	70	−253	455	500	2510	120	8.5	3.0	34.1	106					4–75	

Table 3. *Cont.*

Fuel	Chemical Formula	ρ_{Liquid} (kg/m^3)	$T_{Boiling}$ (°C)	Latent Heat of Vaporization (kJ/kg)	$T_{ignition}$ (°C)	$T_{adiabatic}$ (°C)	LHV (MJ/kg)	LHV Liquid (MJ/L)	HHV Mixture (kJ/L)	A/F_{stoich}	RON	CN	Viscosity (@40 °C) (cSt)	Flash Point (°C)	Reid Vapour Pressure (@38 °C) (kPa)	Flammability Limits (% vol.)	O_2 (%)
Isopropanol	C_3H_7OH	790	82	740			30.3	24		10.3	106			12			
LPG (95% propane)		540	−43	425	455	1980	46	25	3.3	15.6	110					2.1–9.5	
MeFo	$C_2H_4O_2$	957	31.5	464	450		15.8			4.64	115			−19	>100	5–20	53.3
Methane	CH_4	500	−162	510	580	1963	50	22	3.1	17	120			−188		5–15	
Methanol	CH_3OH	800	65	1100	470	1950	19.7	16	3.5	6.4	115	5	0.75	11	32	7–36	50
MTBE	CH_3O C_4H_9	740	55	340		1990	35.2	26		11.7	117			−11		2.4–8	18
Natural Gas (CNG, LNG)		500	−162	419	580	1954	47	21	3.1	17	120					5–15	
Nitromethane	CH_3NO_2	1130	100	600	500	2272	11	12	8.7	1.7							52.5
PL		918							40.5			17.6	2.5				
Petrol RON98	$C_nH_{1.87n}$	720–780	25–210	~350	~260	1995	44	33	3.8	~14.5	98	13–17	0.7	−40	55–100	1.3–8	
$PODE_{3-4}$ or OME_{3-4}	$CH_3O(CH_2O)_nCH_3$	1019					18	19.1				71					
Propane	C_3H_8	520	−43	437	455	1977	46.1	24	3.3	15.7	112			−104	1170	2.1–9.5	
TBA	C_4H_9OH	790	83	570			32.5	26		11.2	113			11		2.4–8	22
Turpentine	$C_{10}H_{16}$	860–900	150–180	285	370		44.4			14.2		20–25	2.5	38	<1	0.7	

Usually the energy content of liquid and solid fuels is characterized by their LHV, as normally the combustion gases are exhausted at temperatures high enough as not to enable condensation. The extremes, in terms of LHV, are carbon (33 MJ/kg, although a common value for coal is 27 MJ/kg) and hydrogen (120 MJ/kg), and in most hydrocarbons their LHV is a function of the H/C ratio. Gaseous fuels such as methane have a high LHV, as its H/C ratio is one of the highest. The energy density of the fuel on a mass base would be an important parameter for mobile applications (namely, aerospace) but not so relevant for stationary applications.

Another important property is the heating value of a stoichiometric air-fuel mixture in terms of volume at atmospheric pressure. This shows the amount of energy that can be introduced into an IC Engine per cycle, which affects torque and power [25]. The importance of this parameter can be illustrated with Hydrogen: although it has the highest HV in terms of mass, its value in terms of volume (of its mixture with air) is one of the lowest (Table 3; Figure 3). Naturally, this parameter affects storage space, which is also critical for mobile applications.

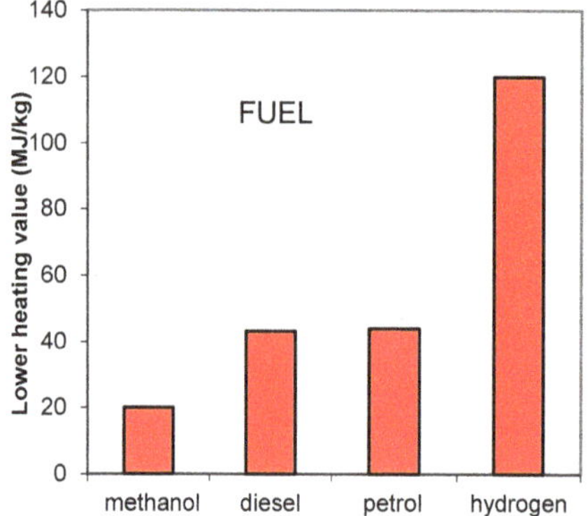

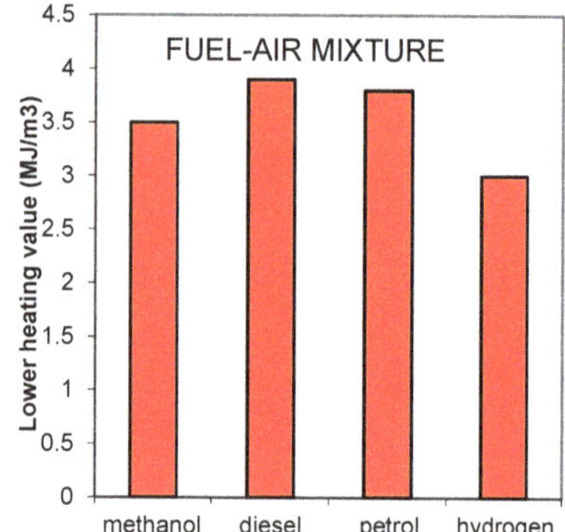

Figure 3. Heating values for the fuel (per mass) and for the stoichiometric air-fuel mixture (per unit volume—[25]).

The latent heat of vaporization is responsible for the cooling effect on the mixture when the fuel is vaporized. Alcohols have a high value, so their mixture with air enters the engine at low temperatures, even when supercharging is used [25].

The value for the stoichiometric air-fuel ratio (A/F) is an indication of the H/C ratio of the hydrocarbon and/or the amount of oxygen (or nitrogen) in its molecule.

The RON (Research Octane Number) and CN (Cetane Number) used to classify commercial petrol and diesel fuels, respectively, are related to the way the fuel self-ignites. High RON numbers are good for Spark Ignition (SI) engines, whereas high CN numbers are good for diesel engines. In fact, these two numbers are opposed [39,40] (Figure 4).

High values for RON indicate a very difficult auto-ignition behaviour, whereas high values for CN specify fuels that auto-ignite easily. Observing Figure 4 it may be seen that there is a clear relation between these numbers. The curve fit yields the following formulae:

$$CN = 56 - 0.39\ RON \tag{1}$$

$$RON = 105 - 0.145\ CN - 0.333\ CN^2 \tag{2}$$

Other physical properties presented in Table 3 are viscosity, flash point and Reid vapour pressure. The viscosity values show the potential for the fuel to be injected as a fine spray, very important for diesel engines. For example, diesel fuel has lower viscosity than biodiesel, so the biodiesel spray is coarser than the conventional diesel spray. On the other hand, ethanol and mainly DME and DEE exhibit much less viscosity, so their injection can be made in fine droplet sprays [41].

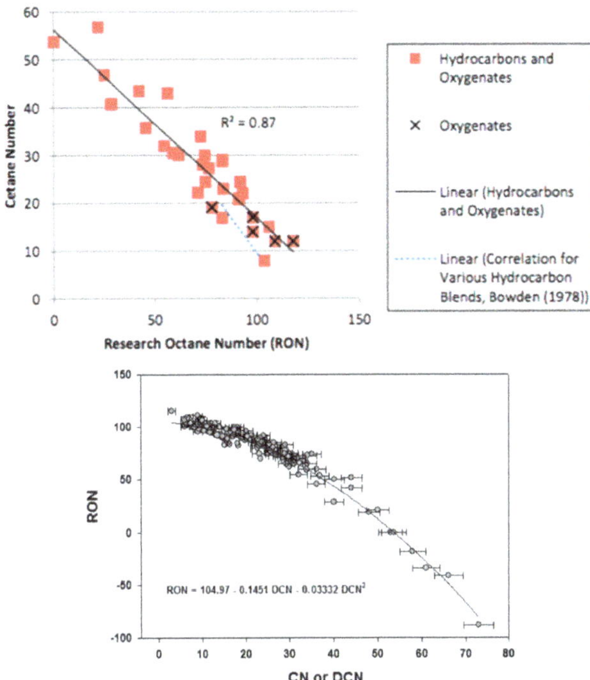

Figure 4. Relation between CN and RON for various oxygenate and hydrocarbon fuels (adapted from [39,40]).

Flash point is the temperature at which the liquid releases enough vapour to produce a stoichiometric mixture with air, therefore sustaining a flame, so it is a property related to safety. If the flash point of a fuel (such as diesel) is well above room temperature, a leak of it will not enable combustion. On the other hand, if the fuel has its flash point below 38 °C (100 °F) it is considered flammable [23].

The Reid vapour pressure is a measure of the volatility of a fuel and is very important for fuels used in SI engines, mainly when they were carburetted fuelled. The flammability limits show the proportions (% in volume) where a spark may ignite the fuel-air mixture. Hydrogen is the fuel with the widest limits for flammability, which is very important to burn very lean mixtures. The burning of very lean mixtures in ICEs has various advantages such as high engine efficiency and low pollutant emissions [25].

Oxygenated fuels, as the name refers, have oxygen in their molecule, so their energy density is reduced by this fact. Half the mass of methanol is oxygen and its LHV is less than half of that of petrol, but as its stoichiometric A/F is also almost half of the petrol, the HV of the air-methanol mixture is, although lower, at about the same level as for petrol. But, as the high latent heat of methanol greatly reduces the air temperature entering the engine, its density is increased by this fact and much more air enters the engine, effectively increasing engine power output (by around 6%, [42]). The fuel with the maximum potential for power boost is nitromethane. Although it has a low HV of 12 MJ/kg, the 1.7 stoichiometric A/F enables a huge amount of fuel to be injected at each cycle, increasing the engine power output to more than 2.3 times the value for petrol [42].

5. Hydrogen

Hydrogen is one of the simplest molecules, with just two atoms joint together, each one with just one proton and an electron. Normally it is in gas form and, unless at very high pressure and/or at very low temperature, its energy density (in terms of volume, or MJ/L–Figure 3) is extremely low (0.11 MJ/L at atmospheric conditions). And this is one of the disadvantages of this fuel: even at very high pressures (750 bar) or very low temperatures (liquefied at 20 K) its energy density is much lower than that of most other liquid fuels (4.7 and 8.6 MJ/L, respectively [43]) And it takes a significant amount of energy to pressurize the hydrogen or to liquefy it, a value that is a significant proportion of its own HV. When compared to other liquid hydrocarbons, one litre of liquid hydrogen actually has less hydrogen (atoms) than a litre of a conventional fuel (and the conventional fuel also has, in addition, carbon atoms).

Some recent developments promise the use of materials to store hydrogen at much lower pressures, such as Liquid Organic Hydrogen Carrier (LOHC) systems [44–46]. However, these methods are complicated, need pressure and/or temperature control and require time for the storage (hydrogenation) and for the recovery (dehydrogenation), often requiring catalysts when liquids are used [47,48]. Yet, these technologies might become viable in the future for specific applications, namely in large-scale stationary cases where the economy of scale eventually compensates for the added complexity.

One other important disadvantage of hydrogen is that its tiny molecule can escape through materials that usually are not permeable to other gases. This requires the use of specific materials for piping and storage, including particular specifications for welding [49].

Although hydrogen can be used in ICEs, its major advantage is to be used, as energy carrier, in fuel cells, where it produces nothing but electricity and water. Hydrogen production from electricity is usually done by water electrolysis in a process that may have an efficiency between 52% and 67%, so 60% seems a good average value. The hydrogen is then used to produce electricity in fuel cells with efficiencies ranging from 50 to 60% (we will use 55%) [18]. Furthermore, it is necessary to store the hydrogen as compressed gas at 350 to 700 bar or as liquid at 20K. This requires 15 MJ/kg for the compression up to 700 bar [50] and 50 MJ/kg for the liquefaction of H_2 [50]. This leads to overall efficiencies of electricity-to-electricity of 29% (compressed H_2) and 19.5% (liquid H_2) when hydrogen is used as an energy carrier. A novel electrolysis process called high temperature electrolysis or steam electrolysis (at 700 to 1000 °C, much higher than the water critical temperature and at high pressures), shows a potential for much higher efficiencies [51].

As said before, petrol and diesel also display more hydrogen content on a volume basis than liquid hydrogen, which makes their synthetic versions also good energy carriers, probably better than hydrogen.

But hydrogen has some important properties to be used in ICEs as an additive. It's very high combustion speed (much higher than petrol) improves the combustion of other fuels even at low fractions (less than 5%). This is beneficial for fuels which burn slowly, such as ammonia [52].

When the percentage of hydrogen is high or when it is burned on its own (pure), the higher adiabatic flame temperature generates a high quantity of NOx, so it is common to inject water as a means to reduce the maximum temperature, especially when supercharging is used. One of the problems of hydrogen is its potential for auto-ignition, as the activation energy (of a spark) required for ignition is very low (0.01 mJ). This fact interferes with the measurement of its knock behaviour and the value for its octane number. The measurement of RON requires an intake temperature of 149 °C, too high for the use of hydrogen. Usually the RON for hydrogen is reported as higher than 100 (Table 3), but some researchers [31] report values as low as 60. Others report RON of over 130 when lean mixtures are used. When the intake is at atmospheric temperature hydrogen shows a very high RON, enabling compression ratios (CR) higher than 14.5:1 without knock, probably helped by its very high combustion velocities [53].

But, as discussed earlier, the major problem with hydrogen is its energy density. 50 L of petrol can be stored in a 72 L tank (weighing 84 kg, including the fuel), whereas the same amount of energy in hydrogen (19 kg) requires a cylindrical tank with 272 L and 129 kg ([54]; see Figure 1). So, the main interests of hydrogen seem to be its potential for use in PEM fuel cells, the absence of CO_2 emissions and its use as an energy carrier and energy storage in stationary applications (although at very high pressures or volumes). If other types of high performance fuel cells capable of consuming liquid fuels are developed, other biofuels (with the potential for not producing fossil CO_2) are used and other synthetic fuels may be used as energy carriers, what will be the benefit of hydrogen? It is our opinion that, if that is the case, the justification for the "hydrogen economy" will lose a lot of its appeal and other high energy density (liquid) synthetic fuels and/or biofuels will likely replace it. So, it seems that the major advantage of hydrogen in transportation is its higher energy density compared to batteries, which makes fuel cell hybrid electric cars a better proposition than full electric vehicles, with higher range and much lower refueling times.

6. Alcohols

The most common alcohol used in propulsion is ethanol, with vast amounts being deployed in Brazil and USA in the so called "flex-fuel" engines. These SI engines can burn petrol, straight ethanol, or any mixture of these two fuels. Although the stoichiometry of both fuels is very different (AFR of 14.5 for petrol and 9.0 for ethanol—see Table 3), the injection system uses the lambda sensor in the exhaust to assess the richness of the mixture and to adjust it. If, for example, the engine is running on straight petrol and the driver fills the tank with ethanol, when the new fuel reaches the injectors, they produce a lean mixture (less fuel than required) but, within one or two seconds the lambda sensor reads the strength of the mixture, sends the information to the ECU (electronic control unit) and the right amount of fuel is then injected to the cylinders and from that point onwards the ECU (electronic control unit) of the engine assumes that fuel. Therefore, only during this very short period the driver may feel some swift glitch in the engine, but then it works seamlessly afterwards.

Methanol is another alcohol occasionally used, mainly in the USA and mainly for racing engines. Ethanol and mostly methanol are exceptional racing fuels for various reasons. They have a high value for RON and a high latent heat of vaporization (see Table 3) leading to cold and dense mixtures entering the engine (more mass) and allowing the use of high compression ratio (CR), which brings higher efficiencies and power [25].

Ethanol is mainly produced from the enzymatic breakdown of starch (grains) leading to sugar and then to ethanol. In the USA the base is corn but in Brazil, where the sugar cane is used, the first transformation is avoided, therefore greatly enhancing the overall efficiency of production.

Methanol is mainly produced from natural gas from the steam reforming equation:

$$CH_4 + H_2O \rightarrow CO + 3H_2 \quad (3)$$

followed by a catalytic reaction between CO and hydrogen:

$$CO + 2H_2 \rightarrow CH_3OH \quad (4)$$

One of the less known advantages of alcohol combustion is the so-called "alcohol bonus". When methanol, is burned, the equation is:

$$CH_3OH + 1.5\ O_2 \rightarrow CO_2 + 2\ H_2O \quad (5)$$

which in terms of moles (which translate into volume) gives:

$$2.5\ V\ (\text{reactants}) \rightarrow 3\ V\ (\text{products}) \quad (6)$$

When petrol is used (considering CH_2):

$$CH_2 + 1.5\ O_2 \rightarrow CO_2 + H_2O \quad (7)$$

which in terms of volumes is:

$$2.5\ \text{V (reactants)} \rightarrow 2\ \text{V (products)} \tag{8}$$

Thus, for the same volume of reactants (2.5) only two volumes are produced when petrol is used, but three are produced when methanol is used. This means there is a substantial higher volumetric expansion when alcohol is used. This can be seen on the indicated diagram, where the methanol shows higher pressure during expansion (Figure 5). This is considering that both petrol and methanol are entirely vaporized when entering the engine, although it is much more difficult to vaporize methanol than petrol, as the latent heat of vaporization of the former (1100 kJ/kg) is much higher than that of the latter (350 kJ/kg), so the relation is even higher.

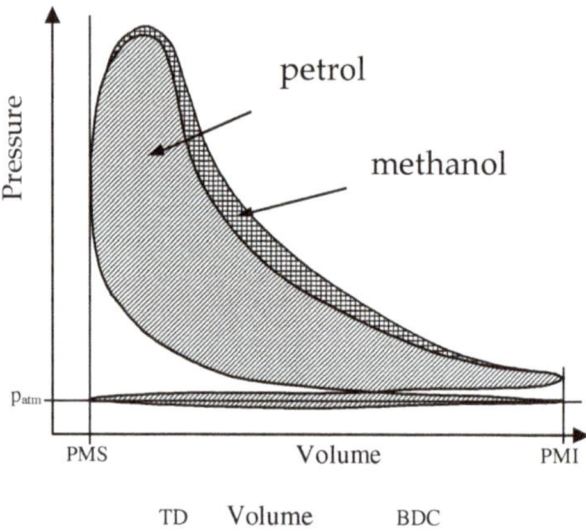

Figure 5. Indicated diagram for petrol and methanol.

As methanol (and ethanol) have much higher latent heat of vaporization than petrol and, for the same power, the amount of injected mass is also considerable higher, the total heat required for the total vaporization of the fuel is much higher when alcohols are used. This creates a cooling effect on the intake mixture, even when supercharging is used. This is beneficial for motorsports, as the thermal loads of engine internals are very high. With the use of alcohols, a supercharged engine can work without intercooling and face no thermal problems or knock.

As the RON values for alcohols are higher than those for petrol (see Table 3), the compression ratio of the engines can be increased without knock, therefore improving power and efficiency. As the adiabatic flame temperature of alcohols is lower than that of petrol, the thermal losses to the combustion chamber are reduced, further improving the overall efficiency.

Other advantages of the alcohols are the fact that, unlike petrol, they mix very well with water, enabling firefighting by the use of water. Throwing water into a petrol fire usually aggravates the problem, as the water is denser than petrol, so the fuel floats over it and spreads easily.

But alcohols also have some problems. The methanol flame has no colour, so it is very difficult to assess whether a fire is taking place. Also, as the heat required for full vaporization of the alcohols is high, mixture preparation may be a problem [55], and a large proportion of liquid may enter the cylinders and "wash" away the oil from the cylinder surfaces, enabling piston-cylinder contact. Also, in cold countries (even in south Brazil) a small tank of petrol is required to start the engine, as ethanol or methanol does not have enough vapour pressure to produce an ignitable mixture. Also, ethanol and especially methanol, tend to induce heavy corrosion on various metals and also other materials such as rubber.

As the flammability limits of alcohols are much wider than those of petrol, the engines may work with much leaner mixtures, improving the efficiency of the engine and reducing all the pollutants.

Despite being a clean and efficient fuel, the major contribution from methanol to transportation is its use to the biodiesel production in the transesterification process [56].

7. Ethers

Ethers are molecules with an atom of oxygen connecting two radicals that usually are similar. For example, dimethyl ether (DME) is composed by two identical methyl radicals connected by the oxygen atom. They are highly flammable liquids or gases which, therefore, may be used in IC engines.

7.1. DME

Dimethyl ether (DME) is the simplest ether and is a gas at atmospheric pressure, but it is easily condensed by applying pressure (<10 bar), so it may be stored in similar containers as propane. Its cetane number (CN > 55) is higher than that of diesel (see Table 3), has a low ignition temperature (320 °C), its viscosity is very low and, as it is composed of 35% oxygen and it has no C-C bonds, its combustion is smoke free [57]. As it is highly volatile, its mixture preparation with air is much easier than diesel, which makes it a perfect compression ignition fuel. Also, it burns fast and without knock (silent combustion—[41]), it has a potential for higher efficiency but it produces more NOx than diesel [56].

DME (and other ethers) can be produced by the dehydration of two alcohol molecules, also producing water (Figure 6). It can also be produced from the "black liquor", a by-product of pulp and paper production, or from lignite-cellulose biomass, which makes it a second-generation biofuel.

$$CH_3OH + CH_3OH \quad \rightarrow \quad CH_3OCH_3 + H_2O$$

Figure 6. Formation of dimethyl ether from methanol.

The compressibility of DME is much higher than diesel, which increases the required energy for fuel compression, although it does not require the huge injection pressures required for fine diesel spray formation. However, it cannot be used directly on diesel injection systems, as it has poor lubrication properties, but a small amount of biodiesel may be added to enable the lubricity. Also, its low density and low heating value require higher injection mass flowrates than diesel but in overall it has the potential for producing more power from the same engine using diesel [56].

7.2. DEE

Diethyl ether (DEE) is a volatile ether conventionally produced as a by-product of ethylene hydration during the production of ethanol, but it can also be produced from the reaction of sulphuric acid with ethanol or by catalytic dehydration of ethanol. As its cetane number is very high (ignition temperature of 160 °C, one of the lowest, Table 3), DEE is used as an IC Engine starting agent, both in SI and diesel engines.

With such a high cetane number (up to 158 [41]) and being able to be stored as a liquid at atmospheric conditions ($T_{boiling}$ = 34 °C, Table 3), this fuel seems to be a good candidate for use in Compression Ignition (CI) engines. It can be added to diesel and, during WWII in Japan, it was used as an additive (up to 5%) in aeroplane engines [41]. Its use as a straight CI engine fuel may have problems as it does not have lubricity and it has tendency to oxidation, producing peroxides. Some organic peroxides are dangerously reactive because they combine both fuel (carbon) and oxygen in the same compound. As it has oxygen in its molecule and has no C-C bonds, its burning does not produce smoke.

8. Esters (Biodiesel)

Common esters are better known as biodiesel and they are good substitutes for fossil diesel fuel in CI engines. They are produced from vegetable oils (and other fats) in esterification or, more commonly, transesterification processes (Figure 7). In the latter process a triglyceride reacts with an alcohol, in the presence of a catalyst, producing the ester and glycerol. Methods for glycerol usage will be discussed in a following chapter.

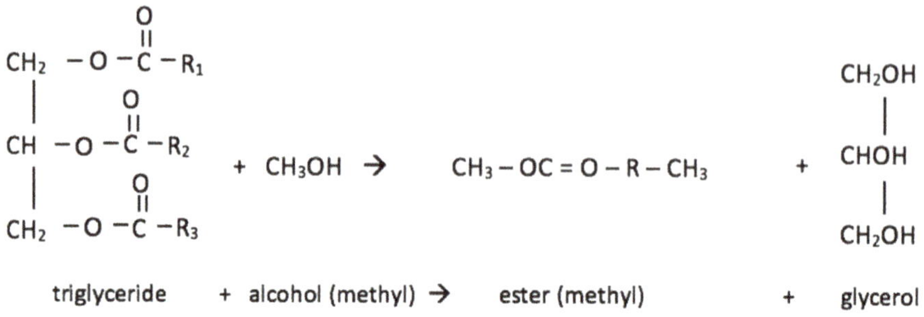

$$\begin{array}{l} CH_2 - O - \overset{O}{\overset{\|}{C}} - R_1 \\ \quad | \\ CH - O - \overset{O}{\overset{\|}{C}} - R_2 \\ \quad | \\ CH_2 - O - \overset{O}{\overset{\|}{C}} - R_3 \end{array} + CH_3OH \rightarrow CH_3 - OC = O - R - CH_3 + \begin{array}{c} CH_2OH \\ | \\ CHOH \\ | \\ CH_2OH \end{array}$$

triglyceride + alcohol (methyl) → ester (methyl) + glycerol

Figure 7. Transesterification process.

The esters of different vegetable oils (rape seed, soy, peanut, sunflower, etc.) are known as biodiesel or FAME (fatty acid methyl ester), if they are produced from methanol. They are usually produced using methanol, but it is possible to produce biodiesel using ethanol. In this case the process is slower and has lower efficiency, but the final product may be considered 100% biofuel, if bioethanol is used and if the vegetable oil is 100% bioproduced. While methanol can also be produced from renewable sources, usually it is derived from natural gas. One of the advantageous properties of biodiesel is its lubricity. When sulphur was removed from diesel, its lubricity was reduced drastically, and the solution was the addition of 2% biodiesel to restore it.

Biodiesel, although with slightly different properties according to the original oil it was made from, is a fuel with higher cetane number than diesel, has no sulphur, the CO and HC emissions are lower [58] and is a biodegradable liquid. In terms of disadvantages, it has higher viscosity than diesel, produces higher values of NOx and is not stable (oxidises) during prolonged storage. Also, the production process is inefficient (is intensive in energy), its heating value is lower than diesel and it may attack elastomers.

For the same amount of injected fuel as diesel, the power reduction using biodiesel should be about 10%, but engine data shows only a reduction of 5% [37], showing a higher efficiency. For the same vehicle energy consumption, the author [58] measured an increase of only 3.5% (in volume) and a reduction of 6% in terms of energy (in fuel) used, when compared to fossil fuel in a long (12,350 km) trip in South America. This also shows a better engine efficiency on the use of biodiesel when compared to diesel [58].

It is important to state that the referred comparative tests were performed in common-rail engines. When using traditional pump-pipe-injector systems the lower compressibility and higher cetane number of the biodiesel generates an earlier and faster combustion which further improves efficiency (and increases NOx) if the engine is developed to minimize NOx emissions. But the lower compressibility of the biodiesel does not interfere with the injection in common-rail systems, so the higher efficiency in these type of engines is only explained by the better combustion potential of the biodiesel [58].

Biodiesel has some disadvantages in relation to diesel. It solidifies at higher temperature (~0 °C) which may be problematic in cold countries. Also, the cold weather additives for diesel do not work for biodiesel, so other additives have to be developed. Biodiesel produced from animal fats (a significant proportion of Brazil biodiesel, 25%) has a much higher solidification temperature (~15 °C) [59].

In prolonged storage it may oxidize and, as it is a biofuel, it may be a source of bacteriologic contamination [59].

Acrolein, which is a toxic substance, is seen as a problem for the biodiesel burning. However, acrolein is a by-product of glycerol burning and biodiesel should have almost no glycerol. In fact, a study [60] showed that biodiesel exhaust emissions may present a lower risk to human health than diesel emissions in IC engines.

There are other processes for the production of biodiesel other than esterification and transesterification. One of the routes involves a mixture of biomass and water (to keep it moisten) undergoing a high temperature (300–350 °C) and high pressure (120–180 bar) process (hydro thermal upgrading—HTU) to remove part (85%) of its oxygen [61]. The resulting oil can be physically or chemically refined into biodiesel. This is a second-generation process.

Iodine Value (IV)

The iodine value (IV) is a measure of the level of unsaturation of the biodiesel or oils. This is an easy test to perform, basically it consists on measuring the amount of iodine that can be added to saturate 100 g of the fuel. The degree of unsaturation relates to the number of double bonds (Figure 8a) between carbon atoms and shows its stability to oxidation and/or to polymerization. A biodiesel from an unsaturated oil has various double bonds. When iodine is added, two atoms are connected to the carbon atoms that were previously connected by the double bond (Figure 8b). The higher the iodine value of the biodiesel (oil or fat), the lower the melting temperature. So, biodiesels produced from animal fat (saturated) have melting points usually above 15 °C [59].

(a) (b)

Figure 8. Partial carbon-hydrogen chain showing a double bond (**a**) and the inclusion of two iodine atoms (**b**).

The hydrogenation of oil occurs when the double bonds of the unsaturated oil are transformed into single bonds and hydrogen atoms are included where the iodine atoms were placed, as seen in Figure 8, right.

A fuel with a higher degree of unsaturation has a higher IV, usually produces higher values of NOx [62–64] and has lower stability to oxidation. With that in mind, in Europe the IV of the biodiesel is limited to 120, restricting biodiesel produced from unsaturated oils such as sunflower or soy [64] and allowing rapeseed oil based biodiesel. This restriction imposes limitations to the biodiesel production of southern European countries and imports from Brazil and the USA (usually from soy) and there is a large debate about it. Biodiesel specifications in the USA, Brazil and Australia do not restrict IV.

Biodiesel produced from animal fats, a highly saturated fat, induces a reduction in NOx [63] and its IV is also low [65]. Brazil uses a mixture of biodiesel produced from soy oil (75%) and tallow fat (25%), enabling the reduction of the high IV from the soy oil biodiesel [65]. However, biodiesel produced from anchovies, an unsaturation fat, has an IV of 185 and tends to reduce the emission of NOx (in 11%), when compared to diesel fuel for similar conditions [66], proving that, at least in some cases, there is no direct link between IV and NOx emissions.

9. Vegetable Oils

The first diesel engines developed by Rudolf Diesel were fuelled by vegetable oils and only later was mineral diesel oil used. One of the problems of using raw vegetable oils is their very high viscosity. It is possible to reduce the viscosity by increasing the temperature of the vegetable oil to values similar to those of diesel prior to injection. But oils and fats have various levels of saturation (double bonds) indicated by their iodine value (IV). The higher the IV, the higher is the probability of the oil or fat to polymerize at high temperature, leading to the formation of heavy and sticky deposits (gums) at the injector tips and piston rings, resulting in a damaged engine.

10. Other Oxygenate Fuels

The previous fuels (alcohols, ethers and esters) have oxygen in their molecule, for which they are called oxygenate fuels. These oxygen atoms significantly improve the fuel combustion and they reduce the potential for particulate matter (PM) production. A work by Harlt [67] showed (Figure 9) that there is a strong correlation between oxygen mass content of the fuel and relative soot (PM) reduction. Furthermore, these researchers proved that oxygenate fuels with higher hydrogen content (higher H/C ratio) tend to further reduce soot formation. Therefore, light oxygenate fuels such as DME or DMM (dimethoxy methane) are better suited as far as soot emissions are concerned. While DME has a high cetane number (~60), DMM, has a relatively low CN of 30 (see Table 3), which reduces the CN number of the mixture diesel-DMM, increasing its ignition delay [68]. But its addition to diesel fuel greatly reduces PM production. DMM has low lubricity so it cannot be used as straight fuel on a diesel engine, without additives for lubricity and cetane number enhancement [67].

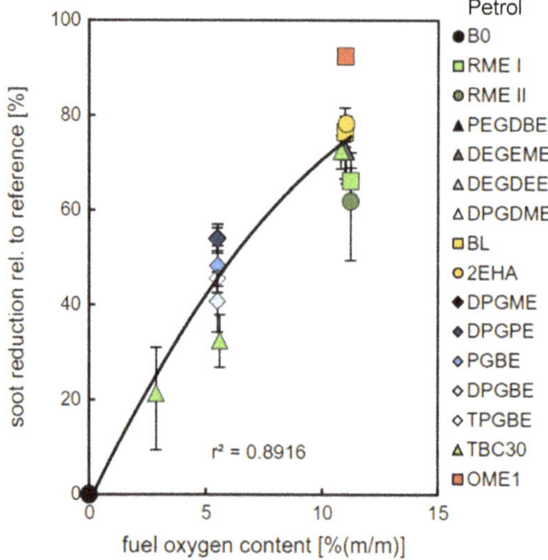

Figure 9. Soot (*PM*) reduction as a function of fuel oxygen content [67].

Diesel engines and recent Spark-Ignition direct injection engines suffer from PM production, as the time for fuel preparation is scarce. In these types of engines, the fuel injection takes place very late in the cycle, reducing the time required for proper fuel-air mixture, leading to PM production. Oxygenate fuels, such as alcohols and ethers, are known for PM production reduction as they lack C-C bonds [69].

Dimethyl carbonate (DMC) and methyl formate (MeFo) are knock resistant (with high RON) esters which are suitable for direct-injection (DI) high-compression Spark-Ignition (SI) engines [70,71]. These fuels also offer the potential for significantly PM production reduction of SI-DI engines, which is a beneficial surplus. They are both used in the chemical industry as solvents and other applications. Both DMC ($C_3H_6O_3$) and MeFo ($C_2H_4O_2$) have equal elemental ratios, so they have the same heating value (Table 3) and similar knocking resistance. Tests comparing to petrol showed a higher fuel consumption for these oxygenates, but the engine power output was significantly increased (by 13%) as a result of a higher mixture heating value when compared to a petrol-air mixture. But the better improvement was in terms of PM reduction, where the particulate number (PN) was reduced by one (DMC) and two (MeFo) orders of magnitude [69] compared to diesel.

In terms of novel oxygenate fuels for compression ignition engines, the oxymethylene ethers OMEs ($CH_3O(CH_2O)nCH_3$) seem very promising [72]. These fuels are also known by the name polyoxymethylene dimethyl ethers (PODEn). This is a class of different fuels, with n varying from 1 (dimethoxy methane or DMM) to more than 5. However, for n = 1 we have the DMM which is very volatile (almost like the DME, which is n = 0) and for n = 2 the fuel has a low flash point [73] so the more usable fuels have the n ranging from 3 to 4 (OME3-4 or PODE3-4). Higher values of n have very high melting points and may precipitate when mixed with diesel fuel. Diesel engine tests of this fuel (PODE3-4 mixed with diesel) showed the potential for a faster combustion, lower PM production and slightly higher NO_x emissions. The engine efficiency improved for all conditions, when compared to straight diesel fuel and the CO and HC emissions were lowered [73]. These fuels have the added benefit of being able to be produced from biomass feedstock [74] and do not have C-C bonds, therefore burning easily and cleanly.

11. Synthetic Fuels—Fischer-Tropsch Process

It is possible to produce synthetic liquid fuels from more traditional fuels such as coal, natural gas or hydrogen. The processes are initiated by the production of syngas (a mixture of hydrogen and carbon monoxide) which then passes through catalytic reactions, such as the Fischer-Tropsch (F-T) process, leading to the production of liquid hydrocarbons [75]. The ratio between H_2 and CO on the syngas and the type of catalyst determines the types of hydrocarbons produced, which can be similar to petrol, diesel or lubricating oil. The relevant equations are the following:

$$nCO + (2n + 1)\,H_2 \rightarrow C_nH_{2n+2} + nH_2O \text{ (paraffins)} \quad (9)$$

$$nCO + 2nH_2 \rightarrow C_nH_{2n} + nH_2O \text{ (olefins)} \quad (10)$$

Diesel F-T has a higher compressibility than fossil diesel (which is not an issue for common-rail engines), has a higher cetane number (Table 3) and the potential for NOx and PM (particulate matter) production is lower [36]. As these synthetic fuels are sulphur free, their combustion is very clean with low PM emitting potential. However, it seems to be sensitive to EGR (exhaust gas recirculation) levels, producing high levels of smoke above a certain value of EGR. But diesel F-T may have different formulations with different distillation curves, which changes some of its properties [76]. These synthetic fuels are hydrocarbons, which do not have oxygen in their molecule, so the reduction of PM production cannot be attributed to that element, as is for oxygenate fuels.

These fuels are seldom known as GTL (gas to liquid). If the base fuel to produce the syngas is biomass, the name changes to BTL (biomass to liquid) and are considered second generation biofuels. During WWII the axis countries had huge shortages of oil, so most of the required fuels and lubricants were produced with these techniques (synthetic fuels) from coal (called CTL—coal to liquid, [77]), like some decades later did South Africa to overcome the oil embargo they were subjected to [78].

11.1. CTL

CTL (coal to liquid) fuels burn cleaner than fossil petrol or fossil diesel, as they are specifically produced to be burned in a particular type of engine. There are two methods to produce them. The indirect coal liquefaction (ICL) requires the crushing of the coal, which then is exposed to high temperature and high pressure together with water (steam) and oxygen to produce the syngas:

$$C + H_2O \rightarrow CO + H_2 \quad (11)$$

and also, with limited air:

$$C + \frac{1}{2}\,O_2 \rightarrow CO \quad (12)$$

Then, the syngas is transformed into liquid fuels by the F-T process. According to the required fuel (petrol, kerosene, diesel or lubricant), the F-T process has to be fed with the right proportions of H_2 and CO, so it is possible to change these proportions, for example by:

$$CO + H_2O + \rightarrow H_2 + CO_2 \quad (13)$$

In the direct coal liquefaction (DCL) the pulverized coal is exposed to hydrogen (hydrogenation) also at high temperatures and pressures (pyrolysis), resulting in a syncrude liquid, which is then refined. The Belgius process involves the mixing of coal with heavy oil recovered from the process and hydrogen at high pressures and temperatures, resulting in a liquid hydrocarbon:

$$nC + (n{+}1)\ H_2 \rightarrow C_nH_{2n+2} \quad (14)$$

The major differences between ICL and DCL are:

- DCL uses just one step and is more energy efficient;
- ICL is easier to control ("design") the type of produced fuel.

The production of huge quantities of CO_2 (almost twice the overall emission of fossil fuels, in a WTW basis, [79]) is one of the major drawback of these processes. Other problems are the large amounts of necessary thermal energy and water. The required water consumption is in the region of $1m^3$ (1 ton) per each barrel of fuel production [77] and the coal (bituminous) consumption is between 0.73 to 1.04 ton per barrel, according to Sasol experience [77]. Suitable catalysts are essential for each process.

11.2. BTL

Some of the processes for transforming solid biomass (plants, wood, crops, straw –lignocellulosic) into liquid fuels (Figure 10) are similar to the above referred processes. The solid biomass is burned under a low oxygen environment (gasification) or is reacted with steam (high pressure and temperature, although lower than CTL) in the presence of adequate catalysts to produce syngas, which is then transformed into liquid fuels using the F-T process. Or, the biomass goes through a pyrolysis process (Figure 10), producing a pyrolysis oil, that is processed and distilled into the required liquid fuels. However, these processes are very biomass intensive (6 ton of biomass to produce 1 ton of BTL, [80]).

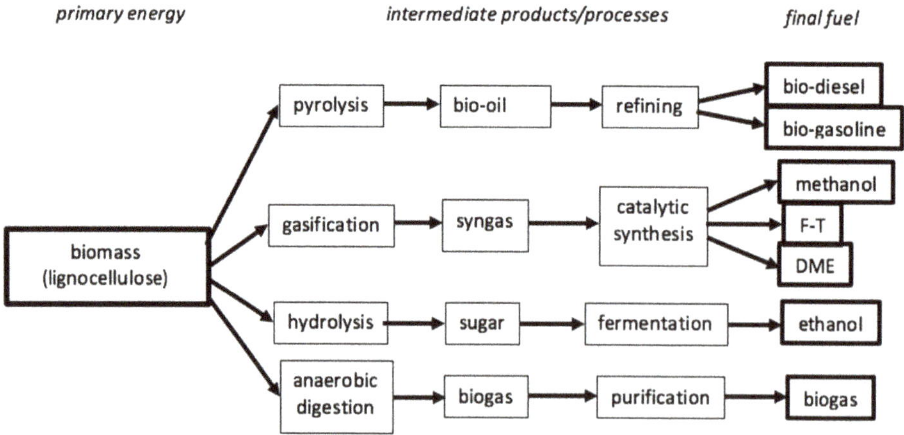

Figure 10. Various processes for the conversion of biomass into engine fuels (BTL).

There are other processes for the transformation of lignocellulosic biomass into fuels such as the second-generation hydrolysis process (Figure 10) followed by the fermentation of the sugar into ethanol and the old process of anaerobic digestion. However, this process requires novel ways of increasing the bioconversion efficiency, such as performing the pre-treatment leading to cell wall degradation [81].

11.3. GTL

The easier way of using the F-T process is using a mixture of natural gas (methane) and steam in a catalyst bed, where it produces the syngas:

$$CH_4 + H_2O + \rightarrow CO + 3H_2 \quad (15)$$

$$CO + H_2O + \rightarrow H_2 + CO_2 \quad (16)$$

These are called gas-to-liquid or GTL. One way to produce BTL fuels is using GTL plants and "hybridizing" them to accept syngas produced from biomass, what is sometimes called hybrid BGTL plants. These plants have a maximum efficiency of around 22% of production fuel derived from biomass [82].

11.4. HVO

Vegetable oils can undergo processes of cracking and/or hydrogenation similar to those in oil refineries, leading to oxygen free linear paraffinic hydrocarbons usually called HVO (hydrogenated vegetable oils) and propane. HVO are not biodiesels, as they have no oxygen in their molecule and have properties similar to fossil diesel.

The hydrogen brakes the links between the glycerol and the fatty acids and deoxidize the hydroxyl and carboxyl groups, leading to a hydrocarbon without oxygen [83]. Part of the carbon is used to "brake" the glycerol and generate propane. So, this process consumes fat and hydrogen and produces long chain hydrocarbons (diesel fuel) using catalysts and high temperature (300 °C) and pressure (50 to 180 bar). CO and CO_2 are also formed as by-products of the process.

The removal of the oxygen alters some of the properties. It reduces the lubricity and its PM (smoke) production and heating value are between those of biodiesel and diesel. The cetane number of HVO is very high (CN = 82: Table 3) which will require a remapping of the engines, namely of the injection advance. Although its heating value is somehow higher than diesel, its lower density results in a lower heating value per unit volume.

11.5. Gasification Fuels—VGO

Gasification fuels can be obtained from biomass using forest residues (small branches and leaves) and black liquor (a paper and pulp production by-product), which produce syngas at high temperature and pressure, followed by the F-T process.

Different plastics (namely those non-recyclable) can undergo a gasification process where, after decontamination, the gases are condensed in a high-quality oil (VGO—vacuum gas oil) that may be distilled into petrol and diesel. Such a processing plant could be a floating platform used to eliminate and treat the enormous amounts of plastic that litter vast parts of the oceans. The process to produce VGO is already used for the recovery of fossil heavy oils, where the heating at low pressure allows the heavy oil to boil at much lower temperatures than at atmospheric pressure. This prevents the production of cokes and therefore increase the liquid fuel production.

11.6. Pyrolysis Fuels (PL)

Pyrolysis is a high temperature reaction without air contact. It produces gases, liquids and solids, being the liquid fractions (PL) the important ones for IC engines. Various substances, such as plastics, biomass and used tires can be used as raw material. Using the latter material (tires), the resulting liquid has a high heating value (over 40 MJ/kg [84]) and some other properties are similar to fossil diesel, but the cetane number is very low (17.6—Table 3), which allows it to be used mixed with fossil diesel (or biodiesel) only in small percentages.

11.7. Electrofuels

The denomination electrofuels has been gaining importance in the last decade [28]. It does mean renewable energy-based fuels which are of non-biologic origin, so they are not based on crops. The EC [85] introduced the term Indirect Land Use Change (ILUC) to account for the consequences (sustainability) of the production of biofuels on the land use. As a sustainability measure, the biofuels produced through ILUC will not be included in terms of renewable targets after 2030. The electro-fuels, such as hydrogen and its derivatives, are basically produced from the (renewable) electricity and water by electrolysis or other physical/chemical processes. Then the hydrogen is combined with the carbon in the CO_2 (through carbon monoxide) to produce the syngas required for the synthesis (Fischer-Tropsch) process [86].

Electro-fuels are very expensive to produce as the energy efficiency of its production is very low, requiring much more energy in the production process than the available in the fuel. Bannon [87] refers overall efficiency values of 73%, 22% and 13% for vehicles running on batteries, on hydrogen fuel-cell and on IC engines burning electrofuel, respectively (Figure 11). So, it is more logical to use the electricity directly stored in batteries in electric vehicles. However, this is not possible in aviation, where the heavy batteries prevent its use. Therefore, airplanes need liquid fuels to fly, so electro-fuels seem to have a future in the huge market for the renewable energy use in the air transport [51,87]. And bear in mind the proposed 50% of EU aviation renewable fuel (electrofuel) by 2050 [85].

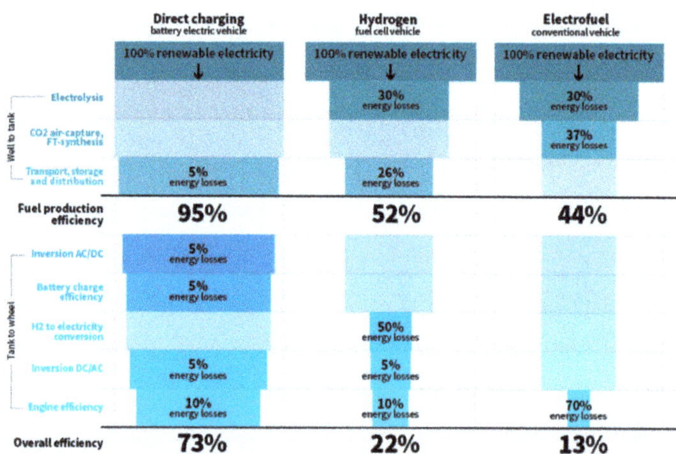

Figure 11. Vehicle overall (WTW) efficiency for electric, fuel cell and electrofuel burning in a IC engine (based on [87]).

When comparing to biofuels, electrofuels from zero-carbon renewable sources (solar, wind, etc.) have much less sustainability risks and use one order of magnitude less land [51] Additionally, the requirement for water is far inferior and there is no risk for groundwater pollution (through chemical fertilization, such as nitrogen). However, if the electricity production is based on any amount of carbon intensity, its inherent low WTW efficiency causes it to produce high levels of CO_2. For example, electrofuels produced from the European energy mix average grid [6] production would have a CO_2 intensity three times higher than current fossil fuels [51]. In terms of cost, the values are extremely high, more than five times the cost of fossil fuels and, therefore, much more than the price for biofuels [88]. One of the important factor on the production of electrofuels is the production of hydrogen by electrolysis, which conventionally has a low efficiency [51]. However, high temperature electrolysis, or steam electrolysis, is a novel process where the hydrogen and oxygen are generated at temperatures between 700 and 1000 °C [89] with much higher efficiencies. The higher efficiencies are, in part, a result of the high temperature of the steam and because the heat required for these high temperatures comes from the subsequent F-T process itself [51]

Another possibility of producing electrofuels is through the high-temperature co-electrolysis of CO_2 and H_2O [90] using solid oxide electrolysis cells (SOECs) [91]. These promising advanced electrochemical energy storage and conversion devices have high conversion efficiencies and convert directly CO_2 and water into syngas, leading directly to the production of F-T fuels, but are still in an early stage of development [92].

11.8. Solar Fuels

Solar fuels are new concepts for renewable liquid fuel production. These fuels are produced from solar energy through direct or indirect techniques. They can be electrofuels, where the required electricity is generated from photovoltaic sources, or fuels generated from processes involving photochemical, thermochemical or biochemical (photosynthesis) involving solar energy [29]. The solar energy in these processes are used for breaking the water and/or the CO_2 producing the H_2 and CO (syngas) required for the subsequent Fischer-Tropsch process. So, some of the earlier mentioned synthetic fuels produced using electricity or thermal energy can also be created using solar energy, through photovoltaic and/or thermal concentration. Enzymatic conversion of CO_2 can also be accomplished with the use of solar energy, leading to chemicals such as methane and CO [93].

One process involves the reaction of CO_2 and H_2O at high temperatures (~1400 °C) in the presence of a cerium oxide catalyst, followed by hydrolysis at 800 °C, producing H_2 and CO [94]. This process of producing liquid hydrocarbons through the processing of the syngas is already being explored commercially, although requires further optimisation involving the use of metal oxides in powerful solar concentrators [29].

Photochemical and photoelectrochemical systems have the active light-absorbing materials directly integrated into the cathode and/or anode electrodes which are placed in contact with the electrolyte. The photosensitized electrodes convert light into an electric current that is then used to split the water into hydrogen and oxygen. The combination of photovoltaic and electrochemical processes is also a promising technology as it allows the separate optimization of both processes [95].

Thermochemical processes have a lot of potential when using very high values of solar concentration [29]. However, the integration of the process into the solar reactor originates significant thermal losses and various other difficulties for upscaling installations have hindered the viability of this approach, with efficiencies lower than 10% [29], prior to considering its use in IC engines, where the efficiency barely reaches the 40% mark or in fuel cells (~60% efficiency).

12. Dimethylfuran (DMF)

Dimethylfuran (DMF—CH_3-C_4H_2-O-CH_3—Table 3) is a biofuel with the potential to substitute petrol, as it has properties between petrol and ethanol. Its RON is even higher than for ethanol, but its knocking behaviour is slightly lower than ethanol, probably because its latent heat of vaporization is much lower (similar to gasoline), preventing the effective mixture cooling of the ethanol [96]. Also, its laminar flame speed is lower than for ethanol, being even slightly lower than for petrol [97]. Its boiling temperature is high (93 °C, Table 3), which makes it a less volatile fuel and more practical for transport and storage, although it may be difficult to start a cold engine. Unlike ethanol and methanol, it is insoluble in water, reducing some of the alcohol's storage problems.

DMF can be obtained from fructose, so it may be a biofuel produced through a chemical or biochemical route using a direct process using catalysts and its production consumes about one third of the energy required for the production of ethanol [98], where its low latent heat of vaporization helps.

13. Nitromethane

Nitromethane (CH_3NO_2—Table 3) is a fuel with oxygen and nitrogen beyond the usual carbon and hydrogen, known for its explosive behaviour and by the huge power improvement it can offer to powerful engines. As it has a high proportion of O and N (52.5% of oxygen and 75.4% of N + O), its heating value is low but, as it has a very low stoichiometric A/F (1.7, Table 3), its mixture with air, in a volume base, carries much more energy than any other fuel, a massive 2.3 times the mixture of air-petrol (see Table 3). The very low A/F requires large amounts of injected fuel (8.5 times more mass, compared to petrol) which, allied to the high latent heat of vaporization (almost twice of the petrol, see Table 3), requires huge amounts of heat to vaporize.

Nitromethane is used mostly on the "top fuel" category of drag racing, where the consumption can be 25 L for a race of 300 m ran in 3.6 s and where the finishing speeds are in excess of 530 km/h. Strangely these 10,000 cv plus engines have no cooling other than the latent heat of the fuel. Some other applications are its use as an additive (~5%), usually of methanol-based fuels, such as those for aero models.

14. Acetylene

This fuel (C_2H_2) was occasionally used in IC engines, mostly during the world wars, as there were no oil-based fuels for the general public. It is produced from the reaction of calcium carbide and water:

$$Ca\,C_2 + 2H_2O \rightarrow C_2H_2 + Ca\,(OH)_2 \tag{17}$$

This process was commonly used in the pit gasometers and on the front lamps of old cars and was occasionally produced in-board (in the trunk of the car) and supplied to the engine.

Acetylene RON is 40 (Table 3), so it is unsuitable for today's high compression ratio (CR) engines, but it has a fast flame propagation speed, which may reduce knock occurrence. One of the best properties of acetylene is its adiabatic temperature, one of the highest, that can exceed the 3000 °C when burned with straight oxygen, but this is not an advantage for an IC engine.

15. Ammonia

Ammonia (NH_3) is an important substance used as a fertilizer throughout the world with a yearly production of over 150 million tons. More than half the world population rely on the enhanced crop production boosted by the nitrogen in the ammonia, but its production is very high energy intensive (uses about 2% of the total energy consumed in the world) and it produces approximately 1% of the CO_2 emissions worldwide. In general, the production of one molecule of NH_3 results in the emission of a molecule of CO_2. In nature, there is the production of ammonia by the decomposition (rotting) of vegetable and animal wastes by bacteria. It can also be produced during the pyrolysis of coal, as a by-product of coke and coal gas production. In these cases, the ammonia appears as ammonium hydroxide, a liquid normally used as a cleaning agent usually known as "ammonia". Ammonia is also used in absorption cycles refrigeration systems.

Anhydrous ammonia (without added water) may be a substitute for petrol in SI engines or even diesel engines and the main interest is to use it as an "energy carrier", to substitute the electricity (or the hydrogen) as a means to transport energy from the place where is produced (wind farm or nuclear power station) to the location where will be used, such as to power vehicles. As there is no carbon in its molecule, it does not produce any CO_2, CO nor HC. Comparing it to hydrogen (another energy carrier), one litre of liquid ammonia (@10 bar and 25 °C) has 30% more hydrogen than 1 L of liquid hydrogen (@-253 °C). Therefore, it makes much more sense to use ammonia as an energy carrier than hydrogen.

In terms of NOx production, the burning of ammonia will produce some thermally (Zeldovich mechanism) but its combustion also produces *NOx* because its molecule has nitrogen atoms. However, ammonia has a relatively low adiabatic temperature, lower than usual hydrocarbons and much lower than hydrogen (Table 3) reducing the potential for Zeldovich NOx production.

In terms of heating value, ammonia has a low value (18.6 MJ/kg, Table 3), less than half of petrol and the comparison gets worse when done in volume, where it has slightly more than 1/3 of the energy of petrol. Also, as ammonia is a gas at atmospheric conditions, it requires a pressure tank, cylindrical or toroidal, similar to those for LPG, where only 80% of the volume can be filled, which reduces the vehicle range.

Other crucial properties of ammonia are the very high auto-ignition temperature (651 °C, Table 3), a very high energy required for ignition (much higher than the required for petrol) and the very high latent heat of vaporization (2450 kJ/kg, compared to 380 for petrol), which introduces further difficulties on its use as an IC engine fuel. Other problems are the narrow limits for ignition (flammability, see Table 3) and the reduced speed for flame propagation, five times slower than petrol [99] and 30 times slower than hydrogen [52]. Adding 4% of ammonia to petrol reduced its burning speed in 15% [99]. As a result, the spark timing required for the burning of the petrol-ammonia mixture increases with the percentage of ammonia (Figure 12, [100]).

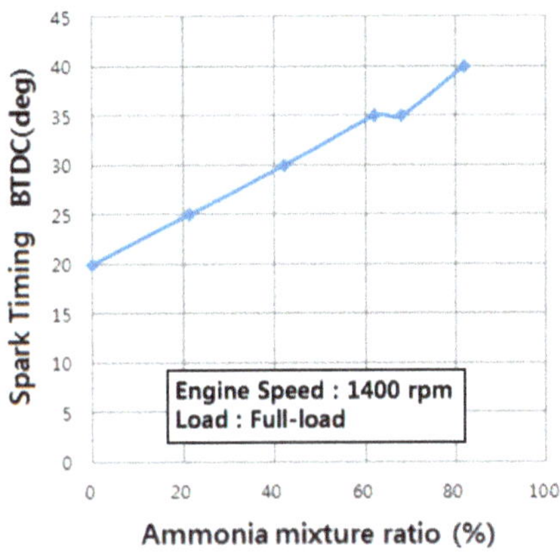

Figure 12. Torque and spark timing required for the mixing of ammonia to petrol (adapted from [100]).

Thus, the combustion of straight ammonia in a SI engine is not easy but it is possible if measures are taken to ensure that the referred combustion deficiencies (very high ignition energy, flammability limits and slow combustion) are overcome, for example using multiple spark location and very high energy ignition and compact combustion chambers [101]. Supercharging seems to be a good option [102] and some researchers used plasma ignition with good results [103]. As one of the major problems is its slow combustion, engine conditions of low charge and high speed can only be achieved using a combustion "promotor" such as hydrogen [53], petrol or even diesel fuel, enabling a better ignibility and increased combustion velocity.

It can be burned in CI engines when mixed with diesel, but it would be advantageous to be mixed with biodiesel or DME as these fuels have higher cetane numbers (CN) [104]. Ammonia causes irritation in small amounts and may be lethal in higher concentrations. However, its distinct and strong smell and the fact that is lighter than air, reduces its risks. Ammonia is largely used in the word, having specific production, storage and delivery installations and procedures, so it has been extensively tested throughout the world. Also, unlike petrol it is not carcinogenic, its combustion does not produce smoke and it is much less prone to explosions [104].

At the moment, ammonia is produced from natural gas (70%) and from coal (30%) by the Haber-Bosch process [105] where hydrogen and nitrogen react ($3H_2 + N_2 \rightarrow 2NH_3$) in an iron oxide catalyst at temperatures ranging from 380 to 500 °C. Ammonia was produced from renewable energy (hydro) by water hydrolyses in the 40s, but high production costs and the aging of the facilities originate production to stop on the 80s [106].

Synthetic gasoline (by the Fisher-Tropsch process) production requires 95.3 MJ/kg and its heating value is 42.5 MJ/kg, which means that it requires 2.25 times its energy content to be produced. In the case of ammonia (Haber-Bosch process plus H_2 production and N_2 separation), the production of 1 kg requires 43.2 MJ [107] and its heating value is 18.6 MJ/kg. Therefore, it requires 2.3 times its energy content to be produced, value similar to the F-T petrol production.

In terms of bio-production, it is necessary to use 2.72 kg of corn to produce 1 kg of ethanol, whereas it is necessary to use 3 kg of the same cereal to produce 1 kg of ammonia, using processes involving gasification and synthesis [108]. As ethanol has an energy density of 25 MJ/kg and ammonia only has 18.6 MJ/kg, in terms of energy, ammonia use is 1.5 worse than ethanol. And the production of ethanol from corn is a bad energetic efficiency when compared to the Brazilian production from sugar cane.

For ammonia to be used as an energy carrier, we know that its production through the Haber-Bosch process ($3H_2 + N_2 \rightarrow 2NH_3$ @ 500 °C and 300 bar) requires 43 MJ/kg [107], already including the production of hydrogen and separation of nitrogen from the air. If ammonia if burned in an IC engine with an efficiency of 40%, then the overall electricity-to-electricity efficiency will be 16%. The same study for liquid hydrogen (see in the hydrogen section) showed this efficiency to be 19.5%.

These examples show that the production and use of ammonia is not, as yet, energetically nor economically viable and it will require the development of more efficient processes. An off-shore wind project for the production of ammonia for use in vessels (Zero Emission Energy Distribution at Sea—ZEEDS [109]) plans water hydrolysis and nitrogen separation from air using the generated wind electricity. It is estimated that this ammonia would be three times more expensive than 3.5% sulphur heavy oil. Another interesting use of ammonia is the direct use in fuel cells [110]), that generally require pure hydrogen.

16. Turpentine

Spirit of turpentine or just turpentine is the resulting fluid from the distillation of pine resin. Traditionally it was used as a solvent (as in dry cleaners) and different trees produce slightly different compositions of the turpentine. Turpentine can also be produced straight from the wood through what is called destructive distillation, a kind of pyrolysis.

Although world annual pine resin production is low at less than a thousand tons [111], there is the development of genetically improved trees to produce high yields of resin [112,113]. However, only the small fraction of 20% of the resin may be transformed into turpentine [114]. Another source of turpentine is the black liquor, a by-product of the pulp and paper industry.

Turpentine ($C_{10}H_{16}$) has been used in IC engines from 1824, when Samuel Morey [115] patented an atmospheric engine using it. But the better-known use of turpentine has been in the post-WWII by Soichiro Honda (founder of the Honda Motor Co), who produced motorcycles with small army-surplus engines (Figure 13) which run on turpentine, that he would sell himself after 1946, because of the lack of petrol in post-war Japan.

Figure 13. Soichiro Honda motorcycle (1947 Honda A-type—adapted from world.honda.com).

Turpentine (see Table 3) has a heating value higher than petrol or diesel and, as its density is also higher, its energy density (in volume or mass) is higher than the conventional fuels. Also, as its stoichiometric A/F (14.2) is lower than petrol, in fact the mixture air-turpentine has more energy than air-petrol and theoretically should produce higher torque and power from the same engine [116]. But its RON is lower than petrol, so the engines require less ignition advance when turpentine is added to petrol.

Turpentine can also be added to diesel fuel, but its low cetane number (20 to 25) tends to reduce the engine efficiency [117], although some authors [118] report a 1–2% efficiency improvement when the mixtures are below 40%. It can be used in dual-fuel mode by fumigation, enabling the substitution of up to 75% of the diesel fuel, with noticeable reductions of smoke production [119].

17. Glycerine (or Glycerol)

Before the intense production of biodiesel, the glycerin was a valuable substance for skin creams, lip sticks and as a food additive. However, the huge quantities of biodiesel produced worldwide generated large surpluses of glycerin, with its value plummeting, as there is no market for it. For each part of generated biodiesel, the transesterification process produces 10% of glycerin.

Although glycerin ($C_3H_8O_3$) may be burned, its atmospheric combustion (at temperatures lower than 300 °C) may produce toxic compounds such as the aldehyde acrolein. Acrolein is produced by the dehydration of glycerol and it is the black and sticky substance produced during the exposure at high temperature of vegetable oils, such as the deposits on the frying pans and responsible for their acrid smell. It has been associated to lung cancer [120], so its emission should be avoided.

As a fuel, glycerin is a very difficult substance to burn in an engine. It solidifies at 18 °C, so it has a high viscosity and has to be injected hot (~100 °C) to enable sufficient atomization. Its auto-ignition temperature is 390 °C, so it is too high for straight use in compression ignition engines. Some researchers mixed it with diesel up to 20% [121], but the intake air needed to be heated up to 100 °C to sustain stable combustion. The power was slightly reduced, and the efficiency was slightly increased, whereas NOx and PM production were reduced, mostly at high power. One of the reported problems was the difficulty to produce and maintain stable mixtures of hydrocarbons and glycerol.

However, at least one company has achieved the combustion of straight glycerin in a diesel engine (Aquafuel Research Ltd., Smarden, Kent, UK). The idea is to increase the intake temperature [122] up to a level that almost any fuel (even petrol) would burn in the diesel engine. So, the intake air should be heated (~200 °C) as well as the fuel (~100 °C), so the glycerin burns cleanly and efficiently. Formula E racing uses electric cars which batteries that have to be charged in the circuit garages. This company developed the electric generators to be used by the different teams to charge the car batteries. These generators are modified diesel engines (Cummins KTA50, 50 L, V16, turbocharged, capable of over 1 MW) that work on glycerol, because it is a clean biofuel. Each generator can produce 850 kWe, enough for charging 40 car batteries in 50 min. These generators are also used by biodiesel producers, enabling them to use the by-product glycerin to produce electricity in their plants. The engines have a drop in power because the intake air density is reduced by the increased temperature. It seems that the high intake temperature is attained by reducing the heat removed in the inter-cooler after the turbo-charger. The emissions NOx and PM were reported as "virtually eliminated" [122], but these statements were reported in the company's site.

A paper reporting on the same company [123] refers 90 °C as the minimum intake temperature to enable stable combustion in a CI engine with glycerin, and 100 °C for petrol RON98. The diesel engine needs to be started on diesel and only can be run on glycerin after warm-up and needs to run again on diesel before being switched-off, to purge the injection system. In terms of laminar flame speed, glycerol is similar to petrol [124].

18. Fage

There are different processes to transform glycerin into usable fuels such as its reaction with dimethyl sulphate and or methanol (producing glycerol dimethoxy ether—[125]), etherification acetylification or anaerobic fermentation [126]. However, these processes are slow and economic unviable. But it is possible to produce FAGE (fatty acid glycerol formal ester) from a reaction between glycerine and other fats (vegetable or animal). FAGE has an LHV lower than biodiesel but is very dense (Table 3), so its energy density (by volume) is similar to biodiesel. But its boiling temperature is almost 300 °C and solidifies at 14 °C, which makes it very viscous.

There is another way of transforming fats without the production of glycerol [126]. Instead of methanol, dimethoxymethane (DMM) is mixed with fat in a combined transesterification-trans-ketalization process producing fatty acid methyl ester (FAME) common biodiesel and FAGE. In overall, the process can be shown as Figure 14).

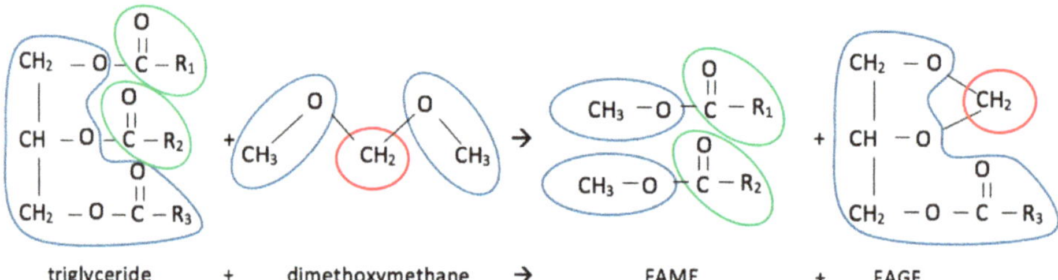

Figure 14. The transesterification-transketalization process of FAME+FAGE production.

19. Conclusions

In a time where the future use of internal combustion engines and/or fossil fuels for road transport is being put into question by many policy makers all over the world, this paper presents an overview of various solutions of alternative fuels that may be used to fuel car engines of the future in a sustainable way.

Some alternatives to the combination internal combustion engine—fossil fuels to propel vehicles exist, such as battery electric cars, fuel cell hybrid vehicles or just conventional vehicles fuelled with renewable and/or biofuels. The latter alternative seems to be particularly attractive, as liquid fuels have very high energy density and they are used in devices (IC engines) that have been developed for over a century. With this in mind, the authors discussed the various propositions for alternatives to fossil fuels. It seems that future liquid fuels will still be burned in IC engines, but the vehicles will be electrically assisted (hybrids) and the exhaust emissions, CO_2 emissions and fuel consumption will be lower than today's fossil fuels.

The various properties, applications and production processes of the various alternative fuels were presented and discussed, from the more conventional alcohols and biodiesel to the more unusual ammonia or turpentine. New concepts such as electrofuels and solar fuels were introduced and discussed. These will be very important in the future, as the land used for the production of biofuels will be limited and restrained. These renewable fuels use significantly less resources in terms of land and water than conventional and second-generation biofuels, but they still have a huge problem of energy efficiency, requiring much more energy for its production than their energy content. These fuels may also be known as "energy carriers" a concept firstly used for hydrogen, whereas they "transport" the energy from where it is produced to where it is used. However, for the most part this is still done with extremely low energy efficiencies.

Some of these fuels are readily available and usable in IC engines with little or no modifications, but others require significant engine modifications and/or adaptations. However, it is possible to burn (in IC engines) unexpected fuels such as ammonia or glycerin, which are currently used in large diesel generators to charge the batteries of Formula E cars. Oxygenated fuels, for instance, are able to retain low particulate matter emissions due to the lack of carbon-carbon bonds.

While synthetic fuels may more easily improve pollutant emissions to the degree that they can be custom manufactured, their greenhouse gas footprint will mostly depend on the sustainability level of its raw materials (fossil/renewable), the energy source and from which they are developed and the energy efficiency of the process.

Author Contributions: Conceptualization, J.M.; methodology, J.M.; software, J.M.; formal analysis, J.M.; investigation, J.M., F.P.B.; resources, J.M., F.P.B.; data curation, J.M., F.P.B.; writing—original draft preparation, J.M., F.P.B.; writing—review and editing, F.P.B., J.M.; visualization, J.M., F.P.B.; supervision, J.M.; project administration, J.M.; funding acquisition, J.M. and F.P.B. All authors have read and agree to the published version of the manuscript.

Acknowledgments: This work has been supported by FCT—Fundação para a Ciência e Tecnologia within the R&D Unit Project Scope UIDB/00319/2020 (MEtRICs—Mechanical Engineering and Resource Sustainability Center).

Conflicts of Interest: The authors declare no conflict of interest.

Glossary

A/F	Air-fuel ratio
BEV	Battery electric vehicle
BTL	Biomass to liquid fuel
CI	Compression ignition
CN	Cetane number
CNG	Compressed natural gas
CO	Carbon monoxide
CO_2	Carbon dioxide
CR	Compression ratio
CTL	Coal to liquid fuel
DCL	Direct coal liquefaction
DEE	Diethyl ether
DMC	Dimethyl carbonate
DME	Dimethyl ether
DMF	Dimethylfuran
DMM	Dimethoxymethane
ETBE	Ethyl *tert*-butyl ether
ECU	Electronic control unit
EGR	Exhaust gas recirculation
FAGE	Fatty acid glycerol formal ester
FAME	Fatty acid methyl ester
F-T	Fischer-Tropsch
GTL	Gas to liquid fuel
HC	Hydrocarbons
HHV	High heating value
HTU	Hydrothermal upgrading
HV	Heating value
HVO	Hydrogenated vegetable oils
H/C	Hydrogen to carbon ratio
IC	Internal combustion
ICE	Internal combustion engine
ICL	Indirect coal liquefaction
IV	Iodine value
LHV	Low heating value
LNG	Liquefied natural gas
LPG	Liquid petroleum gases

MeFo	Methyl formate
MTBE	Methyl *tert*-butyl ether
NOx	Oxides of nitrogen
OEM	Original equipment manufacturer
PL	Pyrolysis fuel liquids
PM	Particulate matter
RON	Research octane number
SI	Spark ignition
TBA	*tert*-Butyl alcohol
USA	United States of America
VGO	Vacuum gas oil
WTW	Well to wheel
WWII	Second World War

References

1. Kumar, R.R.; Alok, K. Adoption of electric vehicle: A literature review and prospects for sustainability. *J. Clean. Prod.* **2020**, *253*, 119911. [CrossRef]
2. Rubens, G.Z.; Noel, L.; Kester, J.; Sovacool, B.K. The market case for electric mobility: Investigating electric vehicle business models for mass adoption. *Energy* **2020**, *194*, 116841. [CrossRef]
3. Biresselioglu, M.E.; Kaplan, M.D.; Yilmaz, B.K. Electric mobility in Europe: A comprehensive review of motivators and barriers in decision making processes. *Transp. Res. Part A* **2019**, *109*, 1–132. [CrossRef]
4. Kalghatgi, G. Is it really the end of internal combustion engines and petroleum in transport? *Appl. Energy* **2018**, *225*, 965–974. [CrossRef]
5. Skrúcaný, T.; Kendra, M.; Stopka, O.; Milojević, S.; Figlus, T.; Csiszár, C. Impact of the Electric Mobility Implementation on the Greenhouse Gases Production in Central European Countries. *Sustainability* **2019**, *11*, 4948. [CrossRef]
6. Zurano-Cervelló, P.; Pozo, P.; Mateo-Sanz, J.M.; Jiménez, L.; Guillén-Gosálbez, G. Sustainability efficiency assessment of the electricity mix of the 28 EU member countries combining data envelopment analysis and optimized projections. *Energy Policy* **2019**, *134*, 110921. [CrossRef]
7. Mossali, E.; Picone, N.; Gentilini, L.; Rodrìguez, O.; Pérez, J.M.; Colledani, M. Lithium-ion batteries towards circular economy: A literature review of opportunities and issues of recycling treatments. *J. Environ. Manag.* **2020**, *264*, 110500. [CrossRef]
8. Zhang, Z.; Fang, W.; Ma, R. Brief review of batteries for XEV applications. *eTransportation* **2019**, *2*, 100032. [CrossRef]
9. Kovács, Z. The End of the Fossil Fuel Car Is on the EU Agenda. Transport & Environment. Available online: https://www.transportenvironment.org/news/end-fossil-fuel-car-eu-agenda (accessed on 5 August 2020).
10. The Climate Center Actions by Countries to Phase out Internal Combustion Engines. Available online: https://theclimatecenter.org/actions-by-countries-phase-out-gas/ (accessed on 5 August 2020).
11. Shaheen, S.; Martin, E.; Totte, H. Zero-emission vehicle exposure within U.S. carsharing fleets and impacts on sentiment toward electric-drive vehicles. *Transp. Policy* **2020**, *85*, A23–A32. [CrossRef]
12. Ribau, J.; Silva, C.; Brito, F.P.; Martins, J. Analysis of Four-Stroke, Wankel, and Microturbine Based Range Extenders for Electric Vehicles. *Energy Convers. Manag.* **2012**, *58*, 120–133. [CrossRef]
13. Volvo Cars Global Newsroom Volvo Cars to Go All Electric. Available online: https://www.media.volvocars.com/global/en-gb/media/pressreleases/210058/volvo-cars-to-go-all-electric (accessed on 5 July 2017).
14. Volvo Cars Global Newsroom Volvo Cars Takes Major Step Towards Its Electrified Future with a Range of New Hybrid Powertrains. Available online: https://www.media.volvocars.com/global/en-gb/media/pressreleases/248975/volvo-cars-takes-major-step-towards-its-electrified-future-with-a-range-of-new-hybrid-powertrains (accessed on 5 August 2020).
15. Sergeev, A. Volvo to Launch Final Generation of Diesel Engines in June. Available online: https://www.motor1.com/news/309040/volvo-last-generation-diesel-engines/ (accessed on 5 August 2020).
16. Volvo Cars Global Newsroom Volvo Cars and Geely Intend to Merge Their Combustion Engine Operations. Available online: https://www.media.volvocars.com/global/en-gb/media/pressreleases/258929/volvo-cars-and-geely-intend-to-merge-their-combustion-engine-operations (accessed on 5 August 2020).

17. International Energy Agency World Energy Outlook. In *Gold Standard of Long-Term Energy Analysis*; IEA/OECD: Paris, France, 2018.
18. Muthukumar, M.; Rengarajan, N.; Velliyangiri, B.; Omprakas, M.A.; Rohit, C.B.; Raja, U.K. The development of fuel cell electric vehicles—A review. *Mater. Today Proc.* **2020**, *3*, 679. [CrossRef]
19. Shusheng, X.; Qiujie, S.; Baosheng, G.; Encong, Z.; Zhankuan, W. Research and development of on-board hydrogen-producing fuel cell vehicles. *Int. J. Hydrog. Energy* **2020**, *4*, 236.
20. Han, X.; Lu, L.; Zheng, Y.; Feng, X.; Li, Z.; Li, J.; Ouyang, M. A review on the key issues of the lithium ion battery degradation among the whole life cycle. *eTransportation* **2019**, *1*, 100005. [CrossRef]
21. Ronanki, D.; Kelkar, A.; Williamson, S.S. Extreme Fast Charging Technology—Prospects to Enhance Sustainable Electric Transportation. *Energies* **2019**, *12*, 3721. [CrossRef]
22. Pavlovic, J.; Tansini, A.; Fontaras, G.; Ciuffo, B.; Otura, M.G.; Trentadue, G.; Bertoa, R.S.; Torino, P. The Impact of WLTP on the Official Fuel Consumption and Electric Range of Plug-in Hybrid Electric Vehicles in Europe. *SAE Tech. Pap.* **2017**, *24*, 0133. [CrossRef]
23. SAE International Hybrid-EV Committee. *J2841: Utility Factor Definitions for Plug-In Hybrid Electric Vehicles Using Travel Survey Data-SAE International*; SAE International Hybrid-EV Committee; SAE International: Warrendale, PA, USA, 2010. [CrossRef]
24. Glassman, I. *Combustion*; Academic Press: New York, NY, USA, 1977.
25. Martins, J. *Motores de Combustão Intern*, 6th ed.; Quantica Editora: Porto, Portugal, 2020; ISBN 9789898927842.
26. Sharma, S.; Panwar, A.K.; Tripathi, M.M. Storage technologies for electric vehicles. *J. Traffic Transp. Eng.* **2020**, *7*, 340–361. [CrossRef]
27. Ferguson, C.R. *Internal Combustion Engines: Applied Thermosciences*; John Wiley: New York, NY, USA, 2001.
28. Bongartz, D.; Doré, L.; Eichler, K.; Grube, T.; Heuser, B.; Hombach, L.E.; Robinius, M.; Pischinger, S.; Stolten, D.; Walther, G.; et al. Comparison of light-duty transportation fuels produced from renewable hydrogen and green carbon dioxide. *Appl. Energy* **2018**, *231*, 757–767. [CrossRef]
29. Mustafa, A.; Lougou, B.G.; Shuai, Y.; Wang, Z.; Tan, H. Current technology development for CO_2 utilization into solar fuels and chemicals: A review. *J. Energy Chem.* **2020**, *49*, 96–123. [CrossRef]
30. Chintala, V. Production, upgradation and utilization of solar assisted pyrolysis fuels from biomass—A technical review. *Renew. Sustain. Energy Rev.* **2018**, *90*, 120–130. [CrossRef]
31. Guíbet, J.-C. (Ed.) *Carburant et Moteurs*; Technip: Paris, France, 1997.
32. Turns, S.R. *An Introduction to Combustion*, 2nd ed.; McGraw-Hill: New York, NY, USA, 2000.
33. Rose, J.W.; Cooper, J.R. *Technical Data on Fuel*, 7th ed.; The British National Committee of the World Energy Conference: London, UK, 1977.
34. The Engineering Toolbox. Available online: https://www.engineeringtoolbox.com (accessed on 5 August 2020).
35. Alma, M.H.; Salan, T. A Review on Novel Bio-Fuel from Turpentine Oil. *Process. Petrochem. Oil Refin.* **2017**, *18*, 1–12.
36. Gill, S.S.; Tsolakis, A.; Dearn, K.D.; Rodriguez-Fernandez, J. Combustion Characteristics and Emissions of Fischer-Tropsch Diesel Fuels in IC Engines. *Prog. Energy Combust. Sci.* **2011**, *37*, 503–523. [CrossRef]
37. Lapuerta, M.; Armas, O.; Rodriguez-Fernandez, J. Effect of Biodiesel Fuels on Diesel Engine Emissions. *Prog. Energy Combust. Sci.* **2008**, *34*, 198–223. [CrossRef]
38. Xu, H.; Wang, C. *A Comprehensive Review of 2,5-Dimethylfuran as a Biofuel Candidate, in Biofuels from Lignocellulosic Biomass: Innovations beyond Bioethanol*; Boot, M., Ed.; Wiley: Weinheim, Germany, 2016; pp. 105–129. ISBN 9783527685318. [CrossRef]
39. Yanowitz, J.; Ratcliff, M.A.; McCormick, R.L.; Taylor, J.D.; Murphy, M.J. *Compendium of Experimental Cetane Numbers*; Technical Report NREL/TP-5400-67585; NREL: Golden, CO, USA, 2017.
40. McCormick, R.; Fiorini, G.; Fouts, L.; Christensen, E.; Yanowitz, J.; Polikarpov, E.; Albrecht, K.; Gaspar, D.J.; Gladden, J.; George, A. Selection Criteria and Screening of Potential Biomass-Derived Streams as Fuel Blendstocks for Advanced Spark-Ignition Engines. *SAE Int. J. Fuels Lubr.* **2017**, *10*, 442–460. [CrossRef]
41. Bailey, B.; Eberhardt, J.; Goguen, S.; Erwin, J. Diethyl Ether (DEE) as a Renewable Diesel Fuel. *SAE Pap.* **1997**, *106*, 1578–1584.
42. Germane, G.J. A Technical Review of Automotive Racing Fuels. *SAE Trans. Fuels Lubr.* **1985**, *94*, 867–878.

43. Do, G.; Preuster, P.; Aslam, R.; Bösmann, A.; Müller, K.; Arltb, W.; Wasserscheid, P. Hydrogenation of the liquid organic hydrogen carrier compound dibenzyltoluene–reaction pathway determination by 1H NMR spectroscopy. *React. Chem. Eng.* **2016**, *3*, 313–320. [CrossRef]
44. Müller, K.; Stark, K.; Emel'yanenko, V.N.; Varfolomeev, M.A.; Zeitsai, D.H.; Shoifetm, E.; Schick, C.; Verevkin, S.P.; Arlt, W. Liquid Organic Hydrogen Carriers: Thermophysical and Thermochemical Studies of Benzyl- and Dibenzyl-toluene Derivatives. *Ind. Eng. Chem. Res.* **2015**, *54*, 7967–7976. [CrossRef]
45. Teichmann, D.; Arlt, W.; Wasserscheid, P.; Freymann, R. A future energy supply based on Liquid Organic Hydrogen Carriers (LOHC). *Energy Environ. Sci.* **2011**, *4*, 2767–2773. [CrossRef]
46. Sotoodeh, F.; Huber, B.J.M.; Smith, K.J. The effect of the N atom on the dehydrogenation of heterocycles used for hydrogen storage. *Appl. Catal. A* **2012**, *419*, 67–72. [CrossRef]
47. Ouma, C.N.M.; Modisha, P.M.; Bessarabov, D. Catalytic dehydrogenation onset of liquid organic hydrogen carrier, perhydro-dibenzyltoluene: The effect of Pd and Pt subsurface configurations. *Comput. Mater. Sci.* **2020**, *172*, 109332. [CrossRef]
48. Jiang, Z.; Gong, X.; Wang, B.; Wu, Z.; Fang, T. A experimental study on the dehydrogenation performance of dodecahydro-N-ethylcarbazole on M/TiO_2 catalysts. *Int. J. Hydrog. Energy* **2019**, *44*, 2951–2959. [CrossRef]
49. Balat, M. Potential importance of hydrogen as a future solution to environmental and transportation problems. *Int. J. Hydrog. Energy* **2008**, *33*, 4013–4029. [CrossRef]
50. Gardiner, M. Energy Requirements for Hydrogen Gas Compression and Liquefaction as Related to Vehicle Storage Needs in DOE Hydrogen and Fuel Cells Program Record. US Department of Energy, 2009. Available online: https://www.hydrogen.energy.gov/pdfs/9013_energy_requirements_for_hydrogen_gas_compression.pdf (accessed on 5 August 2020).
51. Malins, C. What Role is There for Electrofuel Technologies in European Transport's Low Carbon Future? Cerulogy Report. 2017. Available online: https://www.transportenvironment.org/sites/te/files/publications/2017_11_Cerulogy_study_What_role_electrofuels_final_0.pdf (accessed on 5 August 2020).
52. Kobayashi, H. Ammonia Combustion for Energy System. Japan-Norway Hydrogen Seminar, Norwegian Embassy: Tokyo, Japan, 28 February 2017. Available online: http://injapan.no/wp-content/uploads/2017/02/13-Prof.-Kobayashi-Ammonia-Combustion.pdf (accessed on 5 August 2020).
53. Lhuillier, C.; Brequigny, P.; Contino, F.; Rousselle, C. Combustion Characteristics of Ammonia in a Modern Spark-Ignition Engine. *SAE Tech.* **2019**, *24*, 0237. [CrossRef]
54. Poulton, M.L. *Alternative Engines for Road Vehicles*; Computational Mechanics Publications; WIT Press: Southampton, UK, 1994; ISBN 978-1-85312-300-9.
55. Martins, J. Heat and Mass Transfer in Intake Systems of Spark Ignition Engines. Ph.D. Thesis, Department of Chemical Engineering, The University of Birmingham, Birmingham, UK, 1989.
56. Wu, J.; Huang, Z.; Qiao, X.; Lu, J.; Zhang, J.; Zhang, L. Study of Combustion and Emission Characteristics of Turbocharged Diesel Engine Fuelled with Dimethylether. *Energy Power Eng. China* **2008**, *2*, 79–85.
57. McCandless, J.; Teng, H.; Schneyer, J.B. Development of a Variable-Displacement, Rail-Pressure Supply Pump for Dimetil Ether. *SAE Tech.* **2000**, *1*, 0687.
58. Martins, J.; Torres, F.; Torres, E.; Pimenta, H.; Ferreira, V. The Use of Biodiesel on the Performance and Emissions of Diesel Engined Vehicles. *SAE Tech.* **2013**, *1*, 1698. [CrossRef]
59. Bergmann, J.C.; Tupinambá, D.D.; Costa, O.Y.A.; Almeida, J.R.M.; Barreto, C.C.; Quirino, B.F. Biodiesel production in Brazil and alternative biomass feedstocks. *Renew. Sustain. Energy Rev.* **2013**, *21*, 411–420. [CrossRef]
60. Swanson, K.; Madden, M.; Ghio, A. Biodiesel exhaust: The need for health effects research. *Environ. Health Perspect.* **2007**, *115*, 496–499. [CrossRef]
61. Goudriaan, F.; Peferoen, D.G.R. Liquid fuels from biomass via a hydrothermal process. *Chem. Eng. Sci.* **1990**, *45*, 2729–2734. [CrossRef]
62. Öner, C.; Altun, S. Biodiesel production from inedible animal tallow and an experimental investigation of its use as alternative fuel in a direct injection diesel engine. *Appl. Energy* **2009**, *86*, 2114–2120. [CrossRef]
63. Wyatt, V.T.; Hess, M.A.; Dunn, R.O.; Foglia, T.A.; Hass, M.J.; Marmer, W.N. Fuel properties and nitrogen oxide emission levels of biodiesel produced from animal fats. *J. Am. Oil Chem. Soc.* **2005**, *82*, 585–591. [CrossRef]
64. Lapuerta, M.; Rodríguez-Fernández, J.; Mora, E.F. Correlation for the estimation of the cetane number of biodiesel fuels and implications on the iodine number. *Energy Policy* **2009**, *37*, 4337–4344. [CrossRef]

65. Feddern, V.; Cunha, A.; Pra, M.C.; Silva, M.L.B.; Nicoloso, R.S.; Higarashi, M.M.; Coldebella, A.; Abreu, P.G. Effects of biodiesel made from swine and chicken fat residues on carbon monoxide, carbon dioxide, and nitrogen oxide emissions. *J. Air Waste Manag. Assoc.* **2017**, *67*, 754–762. [CrossRef] [PubMed]
66. Altun, S. Effect of the degree of unsaturation of biodiesel fuels on the exhaust emissions of a diesel power generator. *Fuel* **2014**, *117*, 450–457. [CrossRef]
67. Härtl, M.; Seidenspinner, P.; Jacob, E.; Wachtmeister, G. Oxygenate screening on a heavy-duty diesel engine and emission characteristics of highly oxygenated oxymethylene ether fuel OME. *Fuel* **2015**, *153*, 328–335. [CrossRef]
68. Ren, Y.; Huang, Z.; Jiang, D.; Liu, L.; Zeng, K.; Liu, B.; Wang, X. Combustion characteristics of a compression-ignition engine fuelled with diesel–dimethoxy methane blends under various fuel injection advance angles. *Appl. Therm. Eng.* **2006**, *26*, 327–337. [CrossRef]
69. Inal, F.; Senkan, S.M. Effects of oxygenate additives on polycyclic aromatic hydro-carbons (PAHS) and soot formation. *Combust. Sci. Technol.* **2002**, *174*, 1–19. [CrossRef]
70. Pacheco, M.A.; Marshall, C.L. Review of dimethylcarbonate (DMC) manufacture and its characteristics as a fuel additive. *Energy Fuels* **1997**, *11*, 2–29. [CrossRef]
71. Mei, D.; Wu, H.; Ren, H.; Hielscher, K.; Baar, R. Combustion cycle-by-cycle variations in a common rail direct injection engine fueled with dimethylcarbonate-diesel blend. *J. Energy Eng.* **2016**, *142*, 04014059. [CrossRef]
72. Härtl, M.; Gaukel, K.; Pélerin, D.; Wachtmeister, G.; Jacob, E. Oxymethylene ether as potentially CO_2-neutral fuel for clean diesel engines: Part1: Engine testing. *MTZ Worldw.* **2017**, *78*, 52–59. [CrossRef]
73. Wang, Z.; Lium, H.M.; Zhangm, J.; Wang, J.; Shuai, S. Performance, combustion and emission characteristics of a diesel engine fueled with polyoxymethylene dimethyl ethers (PODE3-4)/ diesel blends. *Energy Procedia* **2015**, *75*, 2337–2344. [CrossRef]
74. Zhang, X.; Kumar, X.; Arnold, U.; Sauer, J. Biomass-derived oxymethylene ethers as diesel additives: A thermodynamic analysis. *Energy Procedia* **2014**, *61*, 1921–1924. [CrossRef]
75. Stranges, A.N. Friedrich Bergius and the Rise of the German Synthetic Fuel Industry. *Isis. Univ. Chicago Press* **1984**, *75*, 643–667. [CrossRef]
76. Kreutz, T.G.; Larson, E.D.; Liu, G.; Williams, R.H. Fischer-Tropsch Fuels from Coal and Biomass. In Proceedings of the 25th Annual International Pittsburgh Coal Conference, Pittsburgh, PA, USA, 7 October 2008.
77. Hook, M.; Kjell, A. A review on coal to liquid fuels and its coal consumption. *Int. J. Energy Res.* **2010**, *34*, 848–864. [CrossRef]
78. Spalding-Fecher, R.; Williams, A.; van Horen, C. Energy and environment in South Africa: Charting a course to sustainability. *Energy Sustain. Dev.* **2000**, *4*, 8–17. [CrossRef]
79. Vallentin, D. Policy drivers and barriers for coal-to-liquids (CTL) technologies in the United States. *Energy Policy* **2008**, *36*, 3198–3211. [CrossRef]
80. Chae, H.-J.; Jeong, K.-E.; Kim, C.-U.; Jeong, S.-Y. Development Status of BTL (Biomass to Liquid). *Tech. J. Energy Eng.* **2007**, *16*, 83–92.
81. Xu, N.; Liu, S.; Xin, F.; Zhou, J.; Jia, H.; Xu, J.; Jiang, M.; Dong, W. Biomethane Production From Lignocellulose: Biomass Recalcitrance and Its Impacts on Anaerobic Digestion. *Front Bioeng Biotechnol.* **2019**, *7*, 191. [CrossRef]
82. Wood, D.A.; Nwaoha, C.; Towler, B.F. Gas-to-liquids (GTL): A review of an industry offering several routes for monetizing natural gas. *J. Nat. Gas Sci. Eng.* **2012**, *9*, 196–201. [CrossRef]
83. Lapuerta, M.; Villajosa, M.; Agudelo, J.R.; Boehman, A.L. Key properties and blending strategies of hydrotreated vegetable oil as biofuel for diesel engines. *Fuel Process. Technol.* **2010**, *92*, 2406–2411. [CrossRef]
84. Martinez, J.D.; Lapuerta, M.; Contreras, R.; Murillo, R.; Garcia, T. Fuel Properties of the Tire Pyrolysis Liquid and its Blends with Diesel Fuel. *Energy Fuels* **2013**, *27*, 3296–3305. [CrossRef]
85. EC Renewable Energy–Recast to 2030 (RED II). Available online: https://ec.europa.eu/jrc/en/jec/renewable-energy-recast-2030-red-ii (accessed on 5 August 2020).
86. Transport and Environment, How to Make the Renewable Energy Directive (RED II) Work for Renewable Electricity in Transport, Transport & Environment, Brussels. 2017. Available online: https://www.transportenvironment.org/publications/how-make-renewable-energy-directive-red-ii-work-renewable-electricity-transport (accessed on 5 August 2020).

87. Bannon, E. E-Fuels to Inefficient and Expensive for Cars and Trucks, but May Be Part of Aviation's Climate Solution–Study, Transport & Environment. 2017. Available online: https://www.transportenvironment.org/press/e-fuels-too-inefficient-and-expensive-cars-and-trucks-may-be-part-aviations-climate-solution-%E2%80%93. (accessed on 5 August 2020).
88. Brynolf, S.; Taljegard, M.; Grahn, M.; Hansson, J. Electrofuels for the transport sector: A review of production costs. *Renew. Sustain. Energy Rev.* **2018**, *81*, 1887–1905. [CrossRef]
89. Valderrama, C. High-Temperature Electrolysis. In *Encyclopedia of Membranes*; Drioli, E., Giorno, L., Eds.; Springer: Berlin/Heidelberg, Germany, 2016.
90. Gunduz, S.; Deka, S.; Ozkan, U. Advances in High-Temperature Electrocatalytic Reduction of CO2 and H2O. *Adv. Catal.* **2018**, *62*, 113–165.
91. Zheng, Y.; Wang, J.; Yu, B.; Zhang, W.; Chen, J.; Qiao, J.; Zhang, J. A review of high temperature co-electrolysis of H2O and CO2 to produce sustainable fuels using solid oxide electrolysis cells (SOECs) Advanced materials and technology. *Chem. Soc. Rev.* **2017**, *46*, 1427–1463. [CrossRef]
92. Elder, R.; Cumming, D.; Mogensen, M. *High Temperature Electrolysis in Carbon Dioxide Utilisation: Closing the Carbon Cycle*; Peter, S., Elsje, A.Q., Katy, A., Eds.; Elsevier: Amsterdam, The Netherlands, 2015; pp. 183–209.
93. Shi, J.; Jiang, Y.; Jiang, Z.; Xueyan-Wang, X.; Wang, X.; Zhang, S.; Hanac, P.; Yang, C. Enzymatic conversion of carbon dioxide. *Chem. Soc. Rev.* **2015**, *44*, 5981–6000. [CrossRef]
94. Abanades, S.; Flamant, G. Thermochemical hydrogen production from a two-step solar-driven water-splitting cycle based on cerium oxides. *Sol. Energy* **2006**, *80*, 1611–1623. [CrossRef]
95. Lourenço, A.C.; Reis-Machado, A.S.; Fortunato, E.; Martins, R.; Mendes, M.J. Sunlight-driven CO_2-to-fuel conversion: Exploring thermal and electrical coupling between photovoltaic and electrochemical systems for optimum solar-methane production. *Mater. Today Energy* **2020**, *17*, 100425. [CrossRef]
96. Rothamer, D.A.; Jennings, J.H. Study of the knocking propensity of 2,5-dimethylfuran–gasoline and ethanol–gasoline blends. *Fuel* **2012**, *98*, 203–212. [CrossRef]
97. Tian, G.; Daniel, R.; Li, H.; Xu, H.; Shuai, S.; Richards, P. Laminar Burning Velocities of 2,5-Dimethylfuran Compared with Ethanol and Gasoline. *Energy Fuels* **2010**, *24*, 3898–3905. [CrossRef]
98. Roman-Leshkov, Y.; Barrett, C.J.; Liu, Z.Y.; Dumesic, J.A. Production of dimethylfuran for liquid fuels from biomass-derived carbohydrates. *Nature* **2007**, *447*, 982–985. [CrossRef]
99. Henshaw, P.F.; D'Andrea, T.; Mann, K.R.C.; Ting, D.S.-K. Premixed Ammonia-Methane-Air Combustion. *Combust. Sci. Technol.* **2005**, *177*, 2151–2170. [CrossRef]
100. Woo, Y.; Yang, J.Y.; Lee, Y.J.; Kier, J.N.K. Recent Progress on the Ammonia-Gasoline and the Ammonia-Diesel Dual Fueled Internal Combustion Engines in Korea. In Proceedings of the NH3 Fuel Conference, Des Moines, IA, USA, 21–24 September 2014.
101. Pearsall, T.J.; Garabedian, C.G. Combustion of Anhydrous Ammonia in Diesel Engines. *SAE Tech.* **1967**, 670947.
102. Cooper, J.R.; Crookes, R.J.; Mozafari, A.; Rose, J.W. Ammonia as Fuel for IC Engine. *Int. Conf. Environ. Pollut.* **1991**, *1*, 242–249.
103. Toyota Asian Industries Ammonia Burning Internal Combustion Engine. Patent EP 2378105 A, EP 2378094 A1, 19 October 2011.
104. Kong, S.C.; Gross, C.W. Performance characteristics of a compression-ignition engine using direct-injection ammonia–DME mixtures. *Fuel* **2012**, *103*, 1069–1079.
105. Bartels, J.R. A Feasibility Study of Implementing Ammonia Economy. Master's Thesis, Iowa State University, Ames, IA, USA, 2008.
106. Holbrook, J.H.; Leighty, W. Renewable Fuels: Manufacturing Ammonia from Hydropower, Hydro Review. 2009. Available online: https://www.hydroreview.com/2009/10/01/renewable-fuels-manufacturing/ (accessed on 5 August 2020).
107. Giddey, S.; Badwal, S.P.S.; Munnings, C.; Dolan, M. Ammonia as a Renewable Energy Transportation Media. *ACS Sustain. Chem. Eng.* **2017**, *5*, 10231–10239. [CrossRef]
108. Schill, S.R. Iowa to Get First Biomass-To-Ammonia Plant, Biomass Magazine. Available online: http://biomassmagazine.com/articles/2613/iowa-to-get-first-biomass-to-ammonia-plant (accessed on 5 August 2020).
109. Jaques, B. Sowing the ZEEDS for Zero Emission Fuels, Seatrade Maritime News. 2019. Available online: https://www.seatrade-maritime.com/europe/sowing-zeeds-zero-emission-fuels (accessed on 19 May 2020).

110. Lan, R.; Tao, S. Ammonia as a suitable fuel for fuels cells. *Front. Energy Res.* **2014**, *2*, 35. [CrossRef]
111. Ukkonen, K.A.; Oy, N. Pine Chemicals–Global View. 2016. Available online: https://cdn.ymaws.com/www.pinechemicals.org/resource/resmgr/2016_ic_-_santiago/2016_IC_Santiago-_Presentations/Monday_Speaker_2-_Keijo_Ukko.pdf (accessed on 5 August 2020).
112. Mewalal, R.; Rai, D.K.; Kainer, D.; Chen, F.; Külheim, C.; Peter, G.F.; Tuskan, G.A. Plant-Derived Terpenes: A Feedstock for Specialty Biofuels. *Trends Biotech.* **2017**, *35*, 3. [CrossRef]
113. ARPA-E Plants Engineered to Replace Oil–PETRO Program Overview. 2017. Available online: https://arpa-e.energy.gov/sites/default/files/documents/files/PETRO_ProgramOverview.pdf. (accessed on 5 August 2020).
114. Coopen, J.J.W.; Hone, G.A. *Gum Naval Stores: Turpentine and Rosin from Pine Resin*; National Resources Institute; Food and Agriculture Organization of the United Nations: Rome, Italy, 1995; Available online: http://www.fao.org/3/v6460e/v6460e.pdf (accessed on 5 August 2020).
115. Maurer, L. The Unsolved Mystery of Samuel Morey. Available online: http://www.physics.wisc.edu/~{}lmaurer/academic/morey/SamuelMorey.html (accessed on 5 August 2020).
116. Yumrutas, R.; Alma, M.H.; Ozcan, H.; Kaska, O. Investigation of purified sulfate turpentine on engine performance and exhaust emission. *Fuel* **2008**, *87*, 252–259. [CrossRef]
117. Nallusamy, S.; Sendilvelan, S.; Bhaskar, K.; Prabu, N.M. Analysis of Performance, Combustion and Emission Characteristics on Biofuel of Novel Pine Oil. *Rasayan J. Chem.* **2017**, *10*, 873–880. [CrossRef]
118. Karthikeyan, R.; Nallusamy, N.; Alagumoorthi, N.; Ilangovan, V. Optimization of Engine Operating Parameters for Turpentine Mixed Diesel Fueled DI Diesel Engine Using Taguchi Method. *Mod. Appl. Sci.* **2019**, *4*, 182–192.
119. Karthikeyana, R.; Mahalakshmi, N.V. Performance and Emission Characteristics of a Turpentine–Diesel Dual Fuel Engine. *Energy* **2007**, *32*, 1202–1209. [CrossRef]
120. Zhang, J.J.; Smith, K.R. Household air pollution from coal and biomass fuels in china: Measurements, health impacts, and interventions. *Environ. Health Perspect* **2007**, *115*, 848–855. [CrossRef] [PubMed]
121. Eaton, S.J.; Harakas, G.N.; Kimball, W.K.; Smith, J.A.; Pilot, K.A.; Kuflik, M.T.; Bullard, J.M. Formulation and Combustion of Glycerol–Diesel Fuel Emulsions. *Energy Fuels* **2014**, *28*, 3940–3947. [CrossRef]
122. Aquafuel Introducing Aquafuel G50–Clean Power for Grid Balancing. Available online: https://www.aquafuelresearch.com/ (accessed on 5 August 2020).
123. McNeil, P.; Day, P.; Sirovsy, F. Glycerine from biodiesel: The perfect diesel fuel. *Process. Saf. Environ. Prot.* **2012**, *90*, 180–188. [CrossRef]
124. Jach, A.; Cieślak, I.; Teodorczyk, A. Investigation of glycerol doping on ignition delay times and laminar burning velocities of gasoline and diesel fuel. *Combust. Engines* **2017**, *169*, 167–175. [CrossRef]
125. Ding, X.; Liu, H.; Yang, Q.; Li, N.; Dong, X.; Wanga, S.; Zhao, X.; Wang, Y. A novel route to synthesis of glycerol dimethyl ether from epichlorohydrin with high selectivity. *Biomass Bioenergy* **2014**, *70*, 400–406. [CrossRef]
126. Lapuerta, M.; Rodriíguez-Fernaández, J.; Garcia-Contreras, R. Effect of a Glycerol-Derived Advanced Biofuel–FAGE–on the Emissions of a Diesel Engine Tested under the New European Driving Cycle. *Energy* **2015**, *93*, 568–579. [CrossRef]

Article

Development and Application of Ion Current/Cylinder Pressure Cooperative Combustion Diagnosis and Control System

Denghao Zhu [1], Jun Deng [1], Jinqiu Wang [1], Shuo Wang [1], Hongyu Zhang [2], Jakob Andert [3] and Liguang Li [1,2,*]

1 School of Automotive Studies, Tongji University, Shanghai 201804, China; tj_zdh@outlook.com (D.Z.); eagledeng@tongji.edu.cn (J.D.); wangjinqiu@tongji.edu.cn (J.W.); 1831669@tongji.edu.cn (S.W.)
2 Chinesisch-Deutsches Hochschulkolleg, Tongji University, Shanghai 201804, China; zhanghongyu324@outlook.com
3 Mechatronic Systems for Combustion Engines, RWTH Aachen University, 52074 Aachen, Germany; andert@vka.rwth-aachen.de
* Correspondence: liguang@tongji.edu.cn

Received: 20 September 2020; Accepted: 27 October 2020; Published: 29 October 2020

Abstract: The application of advanced technologies for engine efficiency improvement and emissions reduction also increase the occurrence possibility of abnormal combustions such as incomplete combustion, misfire, knock or pre-ignition. Novel promising combustion modes, which are basically dominated by chemical reaction kinetics show a major difficulty in combustion control. The challenge in precise combustion control is hard to overcome by the traditional engine map-based control method because it cannot monitor the combustion state of each cycle, hence, real-time cycle-resolved in-cylinder combustion diagnosis and control are required. In the past, cylinder pressure and ion current sensors, as the two most commonly used sensors for in-cylinder combustion diagnosis and control, have enjoyed a seemingly competitive relationship, so all related researches only use one of the sensors. However, these two sensors have their own unique features. In this study, the idea is to combine the information obtained from both sensors. At first, two kinds of ion current detection system are comprehensively introduced and compared at the hardware level and signal level. The most promising variant (the DC-Power ion current detection system) is selected for the subsequent experiments. Then, the concept of ion current/cylinder pressure cooperative combustion diagnosis and control system is illustrated and implemented on the engine prototyping control unit. One application case of employing this system for homogenous charge compression ignition abnormal combustion control and its stability improvement is introduced. The results show that a combination of ion current and cylinder pressure signals can provide richer and also necessary information for combustion control. Finally, ion current and cylinder pressure signals are employed as inputs of artificial neural network (ANN) models for combustion prediction. The results show that the combustion prediction performance is better when the inputs are a combination of both signals, instead of using only one of them. This offline analysis proves the feasibility of using an ANN-based model whose inputs are a combination of ion current and pressure signals for better prediction accuracy.

Keywords: Ion current; cylinder pressure; cooperative combustion diagnosis and control; field-programmable gate array; artificial neural network

---

1. Introduction

Recently, internal combustion (IC) engines have been facing the problem of greenhouse gas emissions and exhaust pollutants. Some proposals have even suggested a ban on the sale of vehicles with combustion engines. However, from the statistical results, IC engines provide about 25% of the world's power, while at the same time they only produce about 10% of the world's greenhouse gas emissions [1,2]. Obviously, IC engines have made a huge contribution to promoting the development of the world. Moreover, due to the insufficient charging piles, short battery life, the recycling of used batteries, and other issues of hybrid or electric vehicles, it is highly probable that the IC engines still have a high share in the market [3].

With the application of advanced technologies such as direct fuel injection [4,5], high compression ratio [6,7], lean-burn [8,9], exhaust gas recirculation (EGR) [10,11], advanced ignition system [12,13], waste heat recovery [14,15], water injection [16,17], or novel combustion modes such as homogenous charge compression ignition (HCCI) [18,19], partial premixed compression ignition (PPCI) [20,21], and reactivity controlled compression ignition (RCCI) [22,23], the thermal efficiency and emissions of IC engines have been improved significantly. But at the same time, advanced technologies also increase the occurrence possibility of abnormal combustions such as incomplete combustion, misfire, knock or pre-ignition, and the difficulty in combustion control especially for novel combustion modes which are basically dominated by chemical reaction kinetics. The challenges in precise combustion control are hard to overcome by the traditional engine map-based control method because it cannot monitor the combustion state in every cycle. Real-time cycle-resolved in-cylinder combustion diagnosis and control will be required for the next generation of IC engines.

The cylinder pressure sensor is the most common means of in-cylinder combustion diagnosis and control. Its typical application scenarios are summarized in Figure 1 [24]. With its high-frequency characteristics, it is able to do knock diagnosis and control. With its low-frequency characteristics, it can be used for closed-loop control of maximum brake torque timing (MBT) and air-fuel ratio (A/F). It can also be applied for misfire diagnosis and transient performance control. The following will give a brief overview of the application of cylinder pressure in these fields.

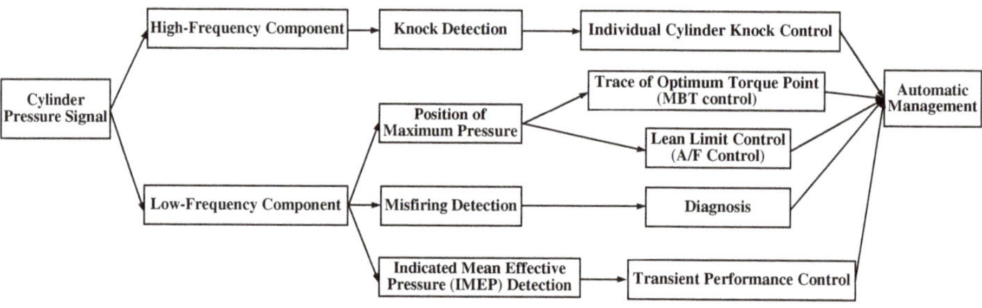

Figure 1. The typical application scenarios of cylinder pressure signals [24].

As early as 1951, Draper et al. [25] proposed that cylinder pressure can be used as a closed-loop control signal for MBT. Subsequently, both Nissan and Honda [24,26] used the cylinder pressure peak time as a characteristic parameter to calculate the MBT compensation for each cycle. With this control strategy, the fuel consumption can be reduced by 1–3%. Zhu et al. [27] compared the performance of using cylinder pressure differential peak, cylinder pressure peak time, and 50% cumulative heat release (CA50) for MBT closed-loop control. It is found that all three parameters can control the ignition timing in the vicinity of MBT, but only the cylinder pressure differential peak does not need to be calibrated, which is more suitable for MBT closed-loop control. For the prediction of the air-fuel ratio based on cylinder pressure, Houpt et al. [28] used the combustion duration to fit the air-fuel ratio, but the limitation of this method is that it is related to the engine operating conditions, fuel, and other parameters. Tunestål et al. [29] employed the heat release rate curve to estimate the air-fuel ratio. However, none of the above methods consider the effect of residual exhaust gas coupling between cycles on the actual air-fuel ratio. Shen et al. [30] considered the transfer of residual exhaust gas between cycles. The results of using cylinder pressure to estimate the air-fuel ratio are very close to the results measured by the oxygen sensor under different operating conditions.

Another common application area of cylinder pressure is combustion diagnosis, especially for abnormal combustion. In 1979, Powell et al. [31] confirmed that the high-frequency oscillation of the cylinder pressure signal can characterize the intensity of knock. Then Sawamoto et al. [32] developed a closed-loop control strategy based on cylinder pressure signals for knock suppression. This strategy successfully expanded the engine torque by 15%. Ravaglioli et al. [33] installed one pressure sensor per cylinder on the Ferrari Formula 1 engines. Based on the combustion information calculated from cylinder pressure, the ignition and injection were adjusted to avoid pre-ignition. Cho et al. [34] improved the evaluation index of knock by reasonable filtering method for cylinder pressure. With this analysis method, it is able to realize transient knock control. Misfire diagnosis is also important as a part of the On-Board Diagnosis II (OBDII) regulation. A sustained misfire will increase the carbon deposit or even damage the three-way catalyst. Shimasaki et al. [35] proposed a misfire detection algorithm based on cylinder pressure when the calculated IMEP is below the predetermined threshold. Similarly, Cesario et al. [36] utilized cylinder pressure for the detection of misfire and partial burning. The misfire fault recognition probability is over 95% at different speeds and loads.

Apart from the cylinder pressure sensor, research on the application of ion current sensor in combustion control has gradually increased in recent years. The basic principle of its formation is that hydrocarbon fuel will generate ions and electrons during the combustion process. When an external electric field is applied to the ion current sensor, the ions and electrons will move directionally to form an ion current. For gasoline engines, the spark plug can be used directly as an ion current sensor, so the cost is much lower than the cylinder pressure sensor. Given its potential in industrial mass production applications, it has been extensively studied in recent years.

Gürbüz [37] studied the correlation between the ion current signal and cylinder pressure in a spark-ignition (SI) engine. A significant positive relationship between periods of combustion, ion current signals, and the local gas temperature was observed. When the engine is running under EGR conditions, the ion current signal is weakened due to the decrease in combustion temperature, but the correlation between the ion current signal and the combustion parameters is still as high as 0.9 [38]. On the natural gas engines and diesel engines, it is also found that the correlation between ion current signal and combustion parameters is higher than 0.9 [39,40]. In addition, on the HCCI engine, the experimental results of Johansson et al. [41] show that the correlation coefficient of the ion current characteristic parameter and CA50 is 0.877, so the ion current signal can be used to estimate the combustion phase of HCCI. Similar results are also reflected in [42,43]. Therefore, under various conditions, the ion current signal has been proved to be highly correlated with combustion parameters, which is the basis for the prediction, diagnosis, and control of combustion.

Hellring et al. [44] took the ion current signal as input and used a neural network model to estimate the CA50 and peak cylinder pressure, which was used for closed-loop control of ignition timing. In addition, similar to the cylinder pressure signal, the ion current signal can also predict the air-fuel ratio, but the difference is that the ion current signal predicts the local air-fuel ratio, that is, the air-fuel ratio near the spark plug. This feature has a special significance for engines that adopt a fuel stratification strategy because the fuel concentration near the spark plug needs to be strictly controlled [45].

Ion current signal can also be used for abnormal combustion diagnosis and control. Auzins et al. [46] tested the success rate of misfire diagnosis in the cases of fuel cutoff and ignition cutoff. Under different operation conditions, the success rate of misfire diagnosis based on ion current signals can be 100%. In our research group, lots of researches have been conducted related to ion current based misfire diagnosis and control [47–49]. Using the amplitude or integral value of ion current signals as the criterion, the misfire can be diagnosed in the current cycle. The methods of re-ignition and re-injection are applied for misfire control. For knock diagnosis and control, Collings et al. [50] compared the experimental results of knock diagnosis with ion current and cylinder pressure and confirmed the feasibility of ion current in knock detection. Laganá et al. [51] studied the characteristics of the ion current signal and knock sensor signal under no-knock, weak knock, and strong knock conditions. When knock occurs, the ion current signal begins to oscillate, and the higher the knocking intensity, the stronger the oscillation. By extracting the frequency domain information of the ion current signal, the correlation coefficient with the knock sensor signal is 0.74. The pre-ignition is a major problem faced by downsizing engines in recent years. Due to the early occurrence of pre-ignition, sufficient time is provided for pressure propagation, which is more destructive to the engine than an ordinary knock. In 2015, for the first time, Tong et al. [52] detected the pre-ignition on a turbocharged gasoline direct injection engine using the ion current signal. Then, in 2019, Wang et al. [53] used ion current signals to detect pre-ignition and used additional fuel injection cooling method to successfully suppress super knock which is induced by pre-ignition under current combustion cycle.

Overall, both the cylinder pressure and the ion current signal can be applied for combustion diagnosis and control. In the past, the two sensors are more like a competing relationship, hence, all related researches only use one of the sensors. However, these two sensors have their own unique features. The cylinder pressure sensor is a "physical sensor" which provides global pressure in the cylinder, while the ion current sensor is a "chemical sensor" that provides localized information around the spark plug. Therefore, in this study, the idea is to combine the information obtained from both sensors to get richer information for combustion diagnosis and control.

At first, two kinds of ion current detection systems are comprehensively introduced and compared. One of them is selected for the subsequent experiments. Then the ion current/cylinder pressure cooperative combustion diagnosis and control system is illustrated and implemented. One application case of using this system for HCCI abnormal combustion control and stability improvement is introduced. Finally, the potential of ion current/cylinder pressure synergy combined with an artificial neural network (ANN) model for combustion prediction has been evaluated.

2. Comparison of Ion Current Detection Systems

Due to the weak ion current signal (microampere level) and the complex electrical environment of the engine, to obtain a high-quality ion current signal requires careful design of the entire system and a large number of experiments for design iteration. In addition, although the basic principle of the circuit is simple, due to the different electronic components and signal processing methods used in each laboratory, the measured ion current signal differs in the waveform, amplitude, and signal-to-noise ratio. In this study, two ion current detection systems are introduced and compared from the hardware level and signal level in both SI and HCCI modes.

One is called direct current power (DC-Power) ion current detection system. Its basic circuit is shown in Figure 2a. On the basis of retaining the original ignition circuit, an ion current detection circuit is connected in parallel. The high voltage power supply module provides DC voltage for the ion current circuit. The capacitor connected in parallel with the high voltage module plays the role of voltage stabilization and energy storage, and its capacitance determines the speed of voltage attenuation. The high voltage silicon stack is used to isolate the instantaneous high voltage generated during ignition so that the electronic components will not be damaged. Another system named the capacitive ion current detection system is shown in Figure 2b. When ignition happens, the discharge current charges the capacitor and is used as the voltage supply to drive the ion current. Since the ion current loop is in series with the ignition loop, various diodes such as transient suppression diodes, fast recovery diodes, Schottky diodes are employed in this circuit to suppress the damage of discharge surge to electronic components. Due to the lack of high voltage modules to continuously charge the capacitor, it is necessary to consider the attenuation of the capacitor voltage. For the engine test, the capacitor is charged once after each cycle of ignition, so the capacitor only needs to ensure sufficient voltage in one cycle. Through calculation, the capacitor charging energy in the circuit is only 0.5 mJ, which is not enough to affect normal ignition process. Besides, the energy consumption of the capacitor in each cycle is only 10% when the data acquisition is completed. Therefore, the voltage decay process of the capacitor can be ignored. The original ion current signal obtained from two detection systems is processed with the same signal processing method including signal differential and resistor-capacitance filtering.

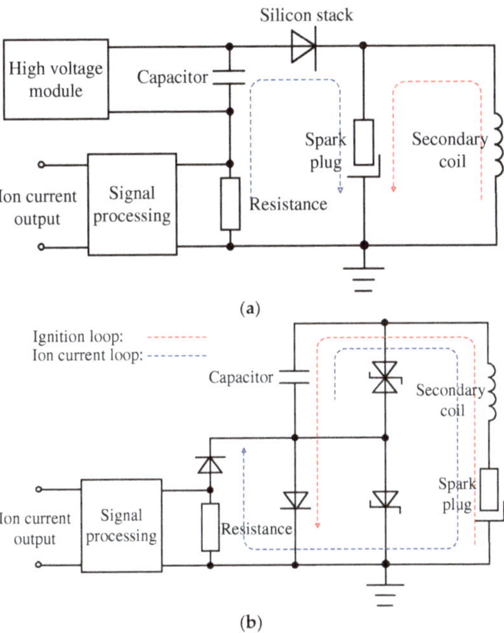

Figure 2. Schematic of two ion current detection systems: (**a**) DC-Power type; (**b**) Capacitive type.

At the hardware level, the advantage of DC-Power ion current detection system is the adjustable output voltage from high voltage power supply. Since the ion current signal is quite sensitive to operating conditions such as air-fuel ratio or intake pressure, its amplitude should be maintained at a level that can be used for combustion analysis. The flexible out voltage can easily meet this requirement. However, the disadvantage is that it is not compatible with mainstream ignition systems, because the DC-Power ion current detection system requires the direction of the ignition current to flow from the center electrode of the spark plug to the side electrode, but the increasingly popular ignition system usually connects the center electrode of the spark plug as "negative electrode", while the side electrode and the whole cylinder are "positive electrode", causing the direction of the ignition current to change from the side electrode to the center electrode. The purpose of this design is mainly to improve the ignition stability, because the temperature of the center electrode is higher than that of the side electrode, and it is easier to emit electrons. Therefore, in order to be able to adopt a DC-Power ion current detection system in the test, the ignition coil of the engine was replaced from the original pen ignition coil (the direction of the ignition current flows from the side electrode to the center electrode) to a relatively traditional static split ignition coil (the direction of the ignition current flows from the center electrode to the side electrode).

For the capacitive ion current detection system, its advantages and disadvantages at the hardware level are just the opposite of DC-Power ion current detection systems. Once the withstand voltage of the capacitor is determined, its output voltage is also fixed. Hence, to change the amplitude of the ion current signal can only be achieved by adjusting the resistance. However, it is found that the amplitude of the ion current signal is not linearly related to the resistance. When the resistance increases to a certain extent such as 5 MΩ, the duration of the ion current signal will also be elongated, causing the signal to deform and fail to truly reflect the combustion process. On the other hand, the capacitive ion current detection system can perfectly adapt to the mainstream ignition systems since the ignition current direction of this system flows from the side electrode to the center electrode as shown in Figure 2b.

In addition to the differences in hardware, there are also differences in the ion current signals measured by the two detection systems. Figure 3 shows the typical ion current signals measured by two ion current detection systems in SI and HCCI modes. In the SI mode, the ion current signal measured by the DC-power ion current detection system has four peaks as shown in Figure 3a, which are an energy storage peak, a discharge peak, a chemical ionization peak, and a thermal ionization peak. Among them, the energy storage peak and the discharge peak are two interference peaks generated on the ion current detection circuit at the moment when the ignition coil starts to store energy and the ignition discharge occurs. When the discharge is over, it begins to enter the main part of the ion current signal. In the early stage of combustion, chemical reactions take place on the flame front, generating a large number of charged particles such as H_3O^+ and electrons. Under the effect of high voltage, a directional movement is generated to form a chemical ionization peak. In the burned area, part of the combustion products will be ionized at high temperatures to generate charged particles such as NO^+ and electrons. This part of the charged particles forms a thermal ionization peak in the middle and late stages of the entire combustion process. Its peak position is very close to the peak position of the cylinder pressure.

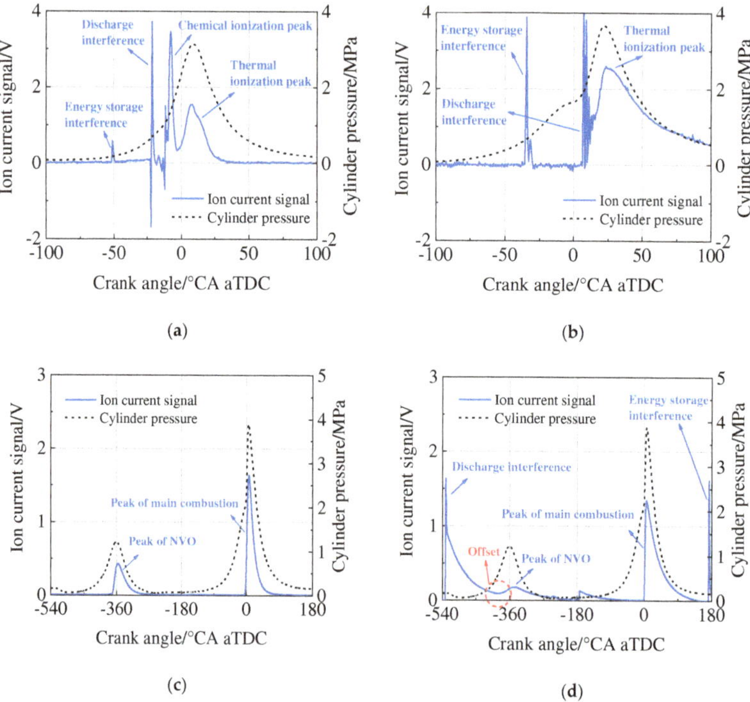

Figure 3. Typical ion current signals measured by two ion current detection systems in SI (1500 r/min, IMEP = 0.28 MPa, Ignition timing = −22 °CA aTDC (after Top Dead Center)) and HCCI modes (1500 r/min, IMEP = 0.3 MPa): (**a**) DC-Power type in SI mode; (**b**) Capacitive type in SI mode; (**c**) DC-Power type in HCCI mode; (**d**) Capacitive type in HCCI mode.

From Figure 3b, energy storage interference and discharge interference also can be observed in the ion current signal measured by the capacitive ion current detection system, but the main part of the ion current has only thermal ionization peaks. Through analysis, the oscillation duration of the discharge interference of this circuit is too long. Because chemical ionization occurs in the early stage of combustion, it is submerged in the discharge peak. This leads to a loss of combustion information at the early stage, which is one of the drawbacks of capacitive ion current detection systems.

In HCCI mode, Figure 3c shows the ion current signal measured by the DC-power ion current detection system. During the negative valve overlap period and the main combustion period, the ion current signal corresponds quite well to the cylinder pressure signal. Since in HCCI mode, ignition is no longer needed, there is no energy storage or discharge interferences. However, for the capacitive ion current detection system, the capacitor in the circuit needs ignition to charge for ensuring normal operation. In order to avoid affecting the combustion process, it can only be ignited once in the exhaust stroke, so it can be seen from Figure 3d that the energy storage and discharge interferences appear at this time. The additional ignition not only causes losses to the ignition system but also affects the ion current signal during NVO. From the red dotted circle in Figure 3d, due to the slower speed of the discharge peak falling to zero, the starting time of the ion current signal during NVO is affected. Moreover, instead of starting from zero, a certain offset can be observed in the ion current signal during NVO. Since the discharge interference varies from cycle to cycle, this offset cannot be quantitatively measured. Therefore, for the HCCI mode, the capacitive ion current detection system is not suitable.

Table 1 summarizes the pros and cons of the two ion current detection systems. Overall, the performance of the DC-power ion current detection system is better than that of the capacitive ion circuit detection system, and it is more suitable for laboratory research. Therefore, in this study, the DC-power ion current detection system is employed in subsequent experiments. The biggest advantage of the capacitive ion current detection system is that it is compatible with mainstream ignition systems, so it is more conducive to industrial applications.

Table 1. Pros and cons of two ion current detection systems.

Ion Current Detection Systems	Pros and Cons	Hardware Level	Signals Level	Summary
DC-Power type	Pros	1. Adjustable output voltage; 2. No additional ignition in HCCI mode.	1. Complete ion current signal in SI mode; 2. No energy storage or discharge interferences in HCCI mode.	More suitable for laboratory research
	Cons	1. Not compatible with mainstream ignition systems.		
Capacitive type	Pros	1. Compatible with mainstream ignition systems.		More conducive to industrial applications
	Cons	1. The output voltage is not adjustable; 2. Additional ignition is required in HCCI mode.	1. Chemical ionization peak is submerged in the discharge oscillation; 2. Ion current signal during NVO is affected by discharge interference in HCCI mode.	

3. Development of Ion Current/Cylinder Pressure Cooperative Combustion Diagnosis and Control System

From the above analysis, an appropriate ion current detection system has been determined. In this chapter, at first, the concept of ion current/cylinder pressure cooperative combustion diagnosis and control will be introduced. Here, the case of HCCI mode is taken as an example, but this concept can also be extended to SI mode. Figure 4 shows the history of ion current and cylinder pressure signals in two consecutive cycles in HCCI mode. The preceding parameters obtained from the previous cycle are suffixed with (i-1), while the parameters obtained from the current cycle are suffixed with (i). Once the signals are acquired in real-time, characteristic parameters can be extracted from either the ion current signal or the cylinder pressure signal. Here, the cycle-resolved combustion control is divided into two categories, feedforward control, and feedback control.

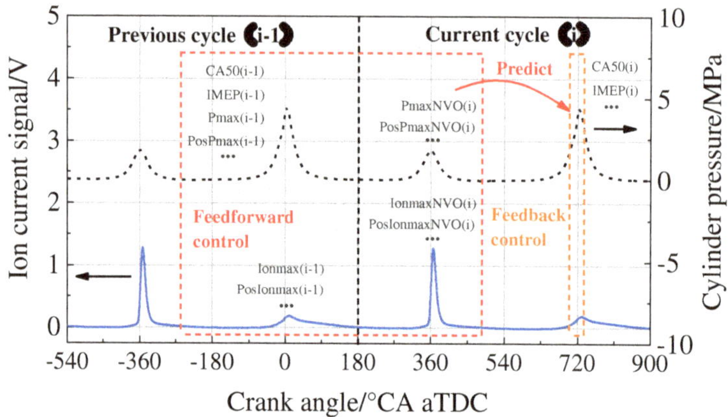

Figure 4. Concept of ion current/cylinder pressure cooperative combustion diagnosis and control (1500 r/min, IMEP = 0.3 MPa).

By extracting parameters of cylinder pressure or ion current from the previous cycle, such as CA50(i-1), indicated mean effective pressure IMEP(i-1), the maximum cylinder pressure Pmax(i-1), the position of maximum cylinder pressure PosPmax(i-1), the maximum ion current signal Ionmax(i-1), the position of maximum ion current signal PosIonmax(i-1), etc., or extracting parameters of cylinder pressure or ion current from current cycle during NVO, such as the maximum cylinder pressure during NVO PmaxNVO(i), the position of maximum cylinder pressure during NVO PosPmaxNVO(i), the maximum ion current signal during NVO IonmaxNVO(i), the position of maximum ion current signal during NVO PosIonmaxNVO(i), etc., these parameters can be applied as input for combustion parameters prediction of the current cycle during the main combustion, such as CA50(i) or IMEP(i). The combustion prediction method could be simple linear regression or highly nonlinear prediction method such as an artificial neural network.

Apart from feedforward control, a combination of ion current and pressure signals can also be employed for feedback control. Here, a type of abnormal combustion, pre-ignition diagnosis, and control in SI mode is taken as an example. Basically, it is difficult to predict pre-ignition with feedforward control. Hence, a more feasible approach is to diagnose pre-ignition as early as possible before its occurrence and then take measures to control it. The whole process including diagnosis and control needs to be finished in dozens of crank angles, which has higher requirements on calculation speed and computing resources of the control system.

The experiments were performed on a modified second generation EA888 engine. Each of the four cylinders is equipped with different compression ratio pistons. Here, only one cylinder with a compression ratio of 16 is used. Specifications of test engine and equipment are shown in Table 2. The original valve system was replaced with a dual UniValve® system both for the intake and exhaust side [54]. A new cylinder head was designed and manufactured to incorporate the dual UniValve® system. Valve lifts can be continuously adjusted from 0 to 8 mm by electrically rotating control shafts, and valve timing can be continuously adjusted within 60 crank angles (CA) by hydraulic driven cam phasers. Besides, valve lifts and valve timing of intake and exhaust side can be adjusted individually. With this advanced system, HCCI combustion can be easily realized through residual gas recirculation without any intake heating assistance. The intake temperature was controlled 25 ± 1 °C, and the coolant temperature was controlled 80 ± 3 °C in SI mode or 90 ± 3 °C in HCCI mode.

Table 2. Specifications of test engine and pressure sensor.

Parameters	Value
Displacement/cm^3	496
Bore/mm	82.5
Stroke/mm	92.8
Compression Ratio	16
Variable Valve Timing/°CA	0–60
Variable Valve Lift/mm	0–8
Fuel	#92 gasoline
Fuel Injection Pressure/MPa	10
Water Injection Pressure/MPa	0.4
Intake Temperature/°C	25 ± 1
Coolant Temperature/°C	SI: 80 ± 3, HCCI: 90 ± 3
Cylinder Pressure Sensor	Kistler 6052C
Charge Amplifier	Kistler SCP2853A

Figure 5 shows the schematic diagram of the engine test bench. The engine is equipped with an intake port water injector and a direct fuel injector. The water rail pressure is 0.4 MPa and the fuel rail pressure is 10 MPa. The spark plug is not only used for ignition, but also an ion current sensor. A side-mounted non-water-cooled pressure sensor is employed to record the in-cylinder pressure trace. The engine prototyping control unit is of type National Instruments® CompactRIO, including a field-programmable gate array (FPGA) module, a real-time controller, and reconfigurable input/output (RIO) modules. The reconfigurable chassis with embedded FPGA is the core of the embedded system. It has an ultra-fast timing resolution of 25 ns. The FPGA module is directly connected to RIO modules, which can access RIO circuits at high speed and flexibly implement functions such as timing, triggering, and synchronization. The real-time controller contains an industrial-grade processor that provides multi-rate control, process execution tracking, on-board data storage, and communication with external devices. The RIO modules contain isolation and conversion circuits, signal conditioning functions, which can be directly connected to industrial sensors or actuators, providing a variety of connection options. This prototyping unit is able to control throttle, ignition coil, injector, UniValve® system, and other actuators.

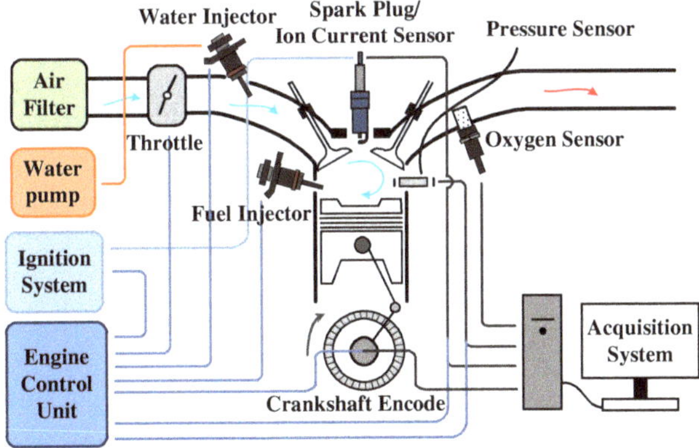

Figure 5. The schematic diagram of engine test bench.

Apart from basic engine control function, the ion current/cylinder pressure cooperative combustion diagnosis and control algorithm are also implemented on CompactRIO as shown in Figure 6. The ion current and cylinder pressure signals are strictly synchronized into the FPGA chassis via the analog input modules. The FPGA box executes high-speed acquisition of two signals and simple parameters calculation, such as the maximum cylinder pressure and its position, the maximum ion current and its position. For HCCI combustion, the characteristic parameters of ion current and cylinder pressure are extracted both in NVO period and main combustion period. These parameters calculated by FPGA will be sent to the real-time controller using FIFO (first-in-first-out) logic. The real-time controller is responsible for complex parameters calculation such as CA50 or IMEP and control algorithm implementation according to specific control targets. The commands that need to be executed will be returned to the FPGA chassis, and then the RIO modules issue instructions to complete the action execution. The whole process will be finished between tens to hundreds of crankshaft angles, depending on in-cycle or cycle-to-cycle control algorithms. At the same time, manipulated variables such as fuel injection, ignition, valve timing, and valve lifts are also acquired by the real-time controller. Under this circumstance, all parameters extracted from ion current or cylinder pressure signals and manipulated variables are strictly guaranteed to be acquired and recorded by the real-time controller in the form of test data exchange stream (TDMS) files synchronously. These data will be used to verify whether the target actuators are executed correctly.

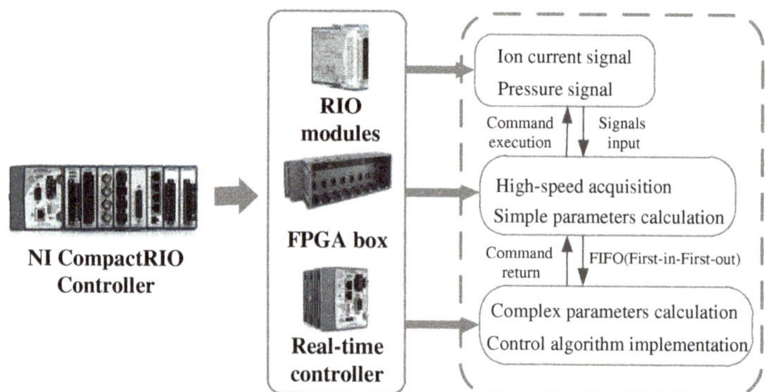

Figure 6. Structure of ion current/cylinder pressure cooperative combustion diagnosis and control system.

Before applying the system for control, online calculation results should be accurate enough. Figure 7 shows the comparison results between online calculation and offline calculation of three randomly chosen parameters calculated from ion current or cylinder pressure signals. They are CA50, IonmaxNVO, and PosPmaxNVO. The raw data of offline calculation is processed by MATLAB®. A correlation analysis is performed to quantitatively evaluate the accuracy of the online calculation. From Figure 7, it can be seen that the correlation coefficient (r) of three parameters are all higher than 0.9, which belongs to "highly relevant". The correlation coefficient between offline calculated PosPmaxNVO and online calculated PosPmaxNVO is up to 0.99. Hence, it can be considered that the online calculation results are accurate enough for combustion diagnosis and control application.

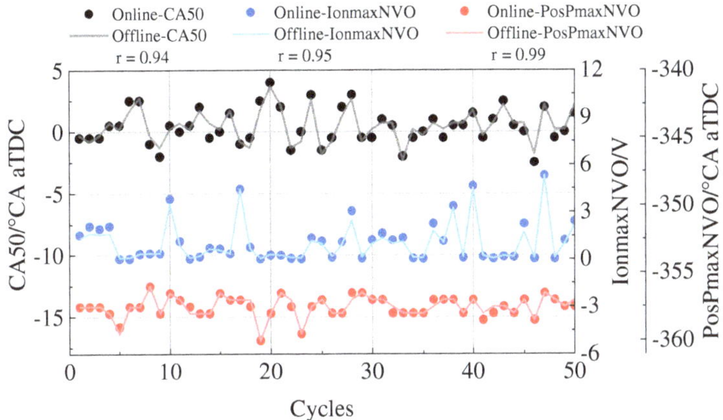

Figure 7. Comparison between online calculation results and offline calculation results.

4. Application of Ion Current/Cylinder Pressure Cooperative Combustion Diagnosis and Control System

After validating the accuracy of online calculation results, in this chapter, an application case of ion current/cylinder pressure cooperative combustion diagnosis and control system in HCCI mode will be introduced.

Figure 8 shows the CA50 and knock intensity (KI) for 400 consecutive cycles of high load boundary conditions in HCCI mode. When the knock intensity is greater than 0.1 MPa, it indicates that knock has occurred. From Figure 8, it can be seen that the knock intensity of the partial cycle is even over 0.4 MPa, which is four times the acceptable limit. Such a strong knock limits the further increases in load. After analysis, knock is divided into two types. The first type is caused by incomplete combustion in the previous cycle, which is the unique regression characteristic of compression ignition engines. The other type is stochastic knock. There is no obvious warning for this kind of knock, so it is difficult to control.

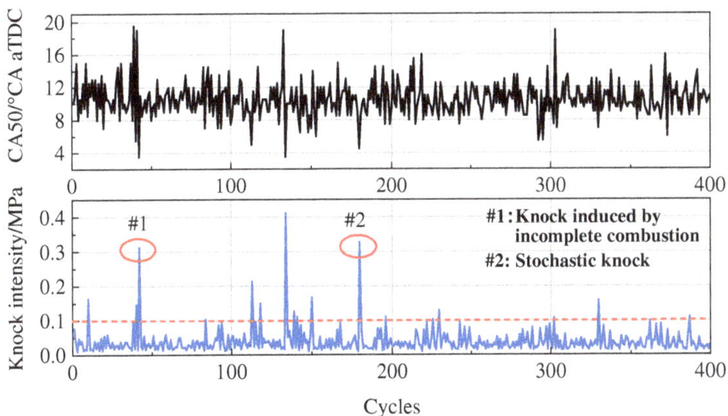

Figure 8. Engine performance at HCCI high load boundary conditions (1500 r/min, IMEP = 0.4 MPa).

For the first type of knock, the reason is apparent. Unburned fuel of incomplete combustion combines with the regularly injected fuel through internal residual gas recirculation, resulting in excess fuel mass, so early combustion or even knock occurs near the top dead center. Thus, this type of knock can be determined by judging whether the previous cycle is incomplete combustion. The left subfigure of Figure 9 shows the CA50 return map at high load boundary conditions. The light grey points are all measurement data, in which the first type of knock cycle is highlighted by blue points. In this area, it is found that linear regression can be performed, and the correlation coefficient is as high as −0.94. It shows that if incomplete combustion occurred in the previous cycle, there will be a great probability of knock in current cycle. This linear regression can be simply fitted by Equation (1).

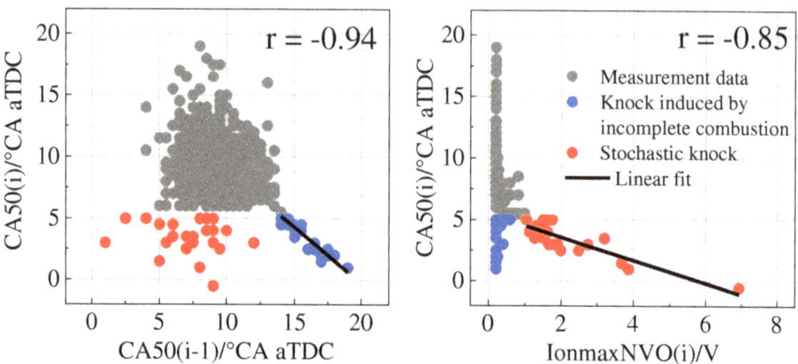

Figure 9. Judgment basis for two kinds of knock.

Except for knock induced by incomplete combustion, stochastic knock highlighted with red points should also be controlled. Since the previous cycle of stochastic knock is normal combustion, the only information available for combustion prediction comes from the negative valve overlap period. By analyzing the correlation between CA50(i) and the cylinder pressure or ion current signal characteristic parameters during NVO, it is found that the IonmaxNVO(i) has a certain corresponding relationship with CA50(i) as shown in the right subfigure of Figure 9. The IonmaxNVO of most cycles is lower than 1 V, but the IonmaxNVO of stochastic cycles is particularly high. Similarly, linear regression is performed for these cycles and the fitted Equation (2) can be obtained. The correlation coefficient is also as high as −0.85, which belongs to highly correlated. This phenomenon confirms that the ion current signal is more sensitive to low-temperature chemical reaction than cylinder pressure signal. When the fuel reforming process during NVO is excessively strong, a stronger ion current signal during NVO can be observed, which becomes an indication of stochastic knock. Compared to the previous results conducted by Wick [16], whose purpose is only to prevent the knock induced by incomplete combustion of HCCI engine with only cylinder pressure for combustion control, the criterions found in this article can predict not only the knock induced by incomplete combustion but also stochastic knock. This result shows the superiority of ion current/cylinder pressure synergy in combustion diagnosis.

$$CA50(i) = -0.96^{*}\ CA50(i\text{-}1) + 18.7 \tag{1}$$

$$CA50(i) = -0.93^{*}\ IonmaxNVO(i) + 5.4 \tag{2}$$

Figure 10 shows the structure of whole control algorithm including knock judgement criterion and knock suprresion method. Real-time CA50 based on cylinder pressure signal and IonmaxNVO based on ion current signal will be caculated in each cycle. The judgment criterions are CA50(i-1) is greater than $CA50_{threshold}$ or IonmaxNVO(i) is greater than $IonmaxNVO_{threshold}$. Knock is consisdered to occur as long as any one of the two judgment criterions is met. It should be noted that the two thresholds need to be calibrated and may be different for different operating conditions or engines. Then CA50 of the next cycle can be linearly estimated by Equation (1) or Equation (2). The predicted CA50 is compared to the target CA50 so that $\Delta CA50_{predict}$ can be obtained. This deviation is the advance of the combustion phase of the current cycle compared to the normal combustion cycle. Once $\Delta CA50_{predict}$ is determined, reasonbale measures should be taken to supress the knock. In this case, intake water injection is employed. The water injection timing is set as −320 °CA aTDC to ensure thorough atomization and evaporation. From the experimental results, the target water injection width $m_{water(i)}$ can be linearly estimated by Equation (3).

$$m_{water(i)} = (\Delta CA50_{predict} + 1.22)/1.4 \tag{3}$$

The controller was experimentally validated at 1500 r/min and 0.4 MPa IMEP. Specifically, the pre-injection and main-injection timing is −400 °CA aTDC and −260 °CA aTDC. The pre-injection and main-injection pulse is 0.37 ms and 0.975 ms. The intake and exhaust valve timing is 15 °CA and 45 °CA. In addition, both the intake and exhaust valve lifts are 3 mm. The test process is to continuously acquired 960 cycles of data, and turn on the control at the 481th cycle to compare the changes in engine combustion before and after the control. Figure 11 shows a comparison of water injection, cylinder pressure, and ion current signals of three continuous cycles before and after control. In the left subgraph, when the controller was off, a typical phenomenon that incomplete combustion and knock occurs alternately can be observed. The 97th cycle is a normal combustion cycle. Then incomplete combustion is randomly appeared in the 98th cycle. Later, knock occurs in the 99th cycle because no control measures are taken. An abnormally high ion current signal can also be observed during NVO in this cycle, which also indicated the occurrence of knock.

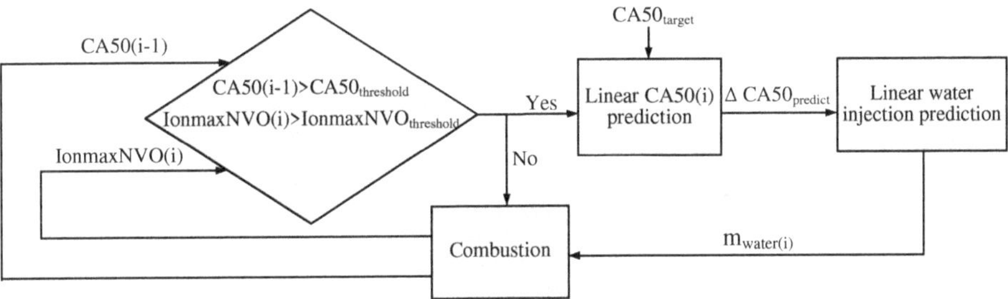

Figure 10. Structure of control algorithm.

After the controller was turn on, the 692nd cycle is a normal combustion cycle. Then in the 693rd cycle, a high ion current during NVO can be observed, which is an indication of knock occurrence. Hence, a 5-V transistor-transistor logic (TTL) water injection signal appears in the 693rd cycle at −320 °CA aTDC, which means that in this cycle water injection was activated, and in this cycle, the knock was successfully suppressed. The followed 694th cycle is also a normal combustion cycle.

To get an overall evaluation of controller performance, the variation of CA50, IMEP, and knock intensity before and after control are shown in Figure 12. The standard deviation of CA50 (σ_{CA50}), the standard deviation of IMEP (σ_{IMEP}), and mean knock intensity are three quantitative evaluation indices of controller performance. In the first 480 cycles, CA50 and IMEP fluctuated due to lacking control measures. The CA50 and IMEP of some cycles significantly deviated from the normal value. Besides, the knock intensity of partial cycles has exceeded the limit of 0.1 MPa. After activating the controller, σ_{CA50} is decreased from 1.88 °CA to 1.75 °CA, σ_{IMEP} is decreased from 0.011 MPa to 0.008 MPa, and mean knock intensity is decreased from 0.048 MPa to 0.041 MPa. More importantly, the knock intensity of controlled cycles can be kept lower than accceptable limit. Therefore, with this algorithm, knock can be effectively suppressed and the combustion stability has been improved. In this case, only 3.5% of cycles need water injection control intervention. Compared to the stationary continuous water injection, intermittent water injection can save water consumption and reduce the possibility of engine parts corrosion and oil emulsification.

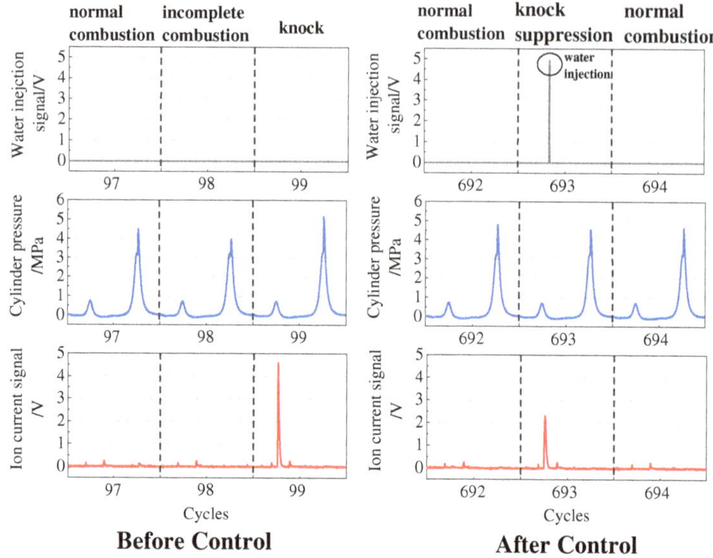

Figure 11. Water injection, cylinder pressure, and ion current signals of three continuous cycles before and after control.

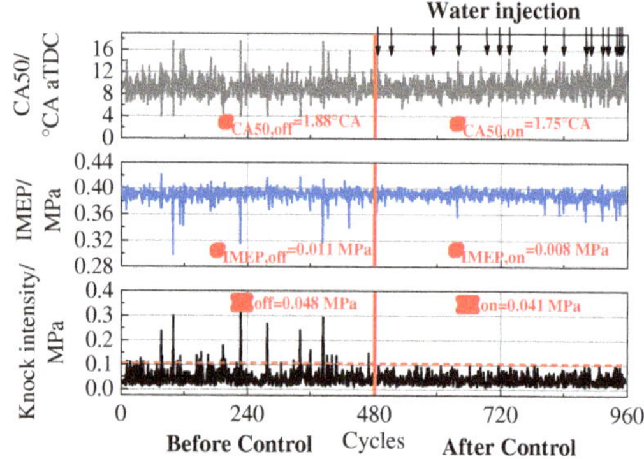

Figure 12. The variation of CA50, IMEP, and knock intensity before and after control.

5. ANN-Based Ion Current/Cylinder Pressures Cooperative Combustion Prediction

In the previous chapter, one application case is introduced. However, the correlation analysis and controller design are mainly based on manual experience and linear regressions. While the engine is an object with multi-factor coupling and strong nonlinear characteristics, considering the advantages of machine learning in dealing with nonlinear problems, in this section, an artificial neural network model is developed to predict the combustion parameters of HCCI mode.

The schematic diagram of an artificial neural network is shown in Figure 13. In order to simply compare the prediction ability of ion current, cylinder pressure, and a combination of the two, the manipulated variables of all measurements were kept unchanged. Therefore, only ion current and cylinder pressure signals make up the data set. Some important characteristic parameters are extracted from both signals as mentioned in the previous session. During pre-processing, both input and observed output variables will be normalized into the range between $(-1, 1)$. In this study, a simple type of artificial neural network, the feedforward neural network (FFNN), is used. Both the input layer and the output layer are one layer. The optimal number of hidden layers and neurons might be different depending on the number of input variables, which is determined using the map sweeping method and will be introduced in detail later. For more intuitive comparison, the results obtained from the output layer need to be anti-normalized. A Levenberg-Marquardt algorithm (trainlm), which combines the advantages of the Newton–Gaussian method and gradient descent method, is utilized as the training algorithm. This method offers faster convergence and lower mean squared error (MSE) than other algorithms when the number of network weights is not too large. A linear function is employed in the output layer. The whole training process is implemented offline using the ANN toolbox supported by MATLAB®.

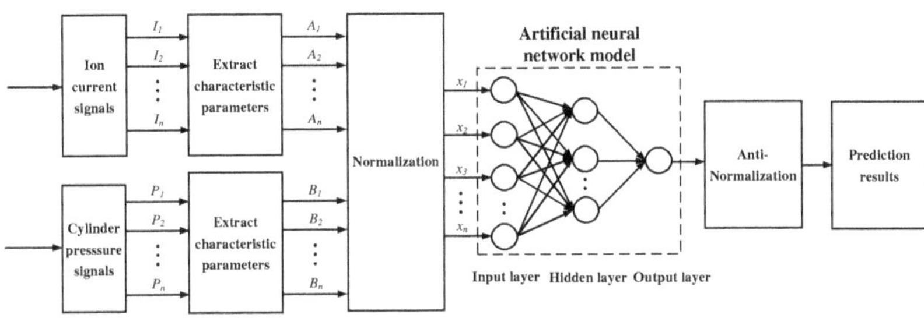

Figure 13. Schematic diagram of artificial neural network.

We chose 154077 valid cycles to train the artificial neural network model. The overall database is divided into three sets. The validation and the testing set each own 15% of the entire database, while the remaining 70% is used as the training set. In order to compare the prediction performance of the model employing only ion current related-parameters, only cylinder pressure-related parameters, or both, three sets of input variables were selected as shown in Table 3. The input variables of the first model named as ANN_{IC} are only ion current-related parameters, including Ionmax(i-1), PosIonmax(i-1), IonmaxNVO(i), PosIonmaxNVO(i). The input variables of the second model named as ANN_{CP} are only cylinder pressure-related parameters, including CA50(i-1), IMEP(i-1), Pmax(i-1), PosPmax(i-1), PmaxNVO(i), and PosPmaxNVO(i). Meanwhile, the third group named ANN_{IP} combines all the mentioned variables as input variables. The output variable is selected as CA50(i), which is identical for all groups.

Table 3. The input and out variables of three ANN models.

Model Name	ANN_{IC}	ANN_{CP}	ANN_{IP}
Input variables	$x_i = \begin{pmatrix} Ionmax(i-1) \\ PosIonmax(i-1) \\ IonmaxNVO(i) \\ PosIonmaxNVO(i) \end{pmatrix}$	$x_i = \begin{pmatrix} CA50(i-1) \\ IMEP(i-1) \\ Pmax(i-1) \\ PosPmax(i-1) \\ PmaxNVO(i) \\ PosPmaxNVO(i) \end{pmatrix}$	$x_i = \begin{pmatrix} Ionmax(i-1) \\ PosIonmax(i-1) \\ CA50(i-1) \\ IMEP(i-1) \\ Pmax(i-1) \\ PosPmax(i-1) \\ IonmaxNVO(i) \\ PosIonmaxNVO(i) \\ PmaxNVO(i) \\ PosPmaxNVO(i) \end{pmatrix}$
Output variable		$y_i = CA50(i)$	

After the input and output variables are determined, the optimal ANN structure for three models needs to be determined. The approach is to do the map sweeping of hidden layers and neurons. Specifically, the hidden layer sweeps from 1 to 4, and the neurons of each hidden layer sweep from 4 to 10. Then each combination will be automatically trained and the MSE can be calculated with Equation (4):

$$MSE = \frac{1}{n}\sum_{i=1}^{n}(y_i - \widetilde{y_i})^2 \tag{4}$$

in which y_i is the observed value of output variables, $\widetilde{y_i}$ is the predicted value of output variables, and n is the number of data points.

The cost function as shown in Equation (5) is the evaluation index to determine the optimal ANN structure, which has compromised the training error, validation error, and training duration [55]. The combination with the lowest cost will be the optimal hidden layer and neurons for a specific model.

$$K = \frac{1}{3}\frac{MSE_{train}}{\max(MSE_{train})} + \frac{1}{2}\frac{MSE_{valid}}{\max(MSE_{valid})} + \frac{1}{6}\frac{t_{train}}{\max(t_{train})} \tag{5}$$

in which MSE_{train} is the MSE of training data set, MSE_{valid} is the MSE of validation data set, and t_{train} is the training duration.

Figure 14 shows the cost map of three ANN models. Basically, the difference in cost among various combinations is small. The cost increases with the increase of hidden layers and neurons since the training duration increases significantly. The minimal cost is highlighted with a star symbol. For the ANN_{IC}, the optimal number of the hidden layer is 1 with 9 neurons per layer. For the ANN_{CP}, the optimal number of the hidden layer is 1 with 5 neurons per layer. And For the ANN_{IP}, the optimal number of the hidden layer is 2 with 6 neurons per layer. After optimal network structures have been settled, each ANN model will be trained five times with randomly generated initial matrices. Because different initial matrixes may lead to different final values of matrixes, the model with the lowest cost will be selected and analyzed.

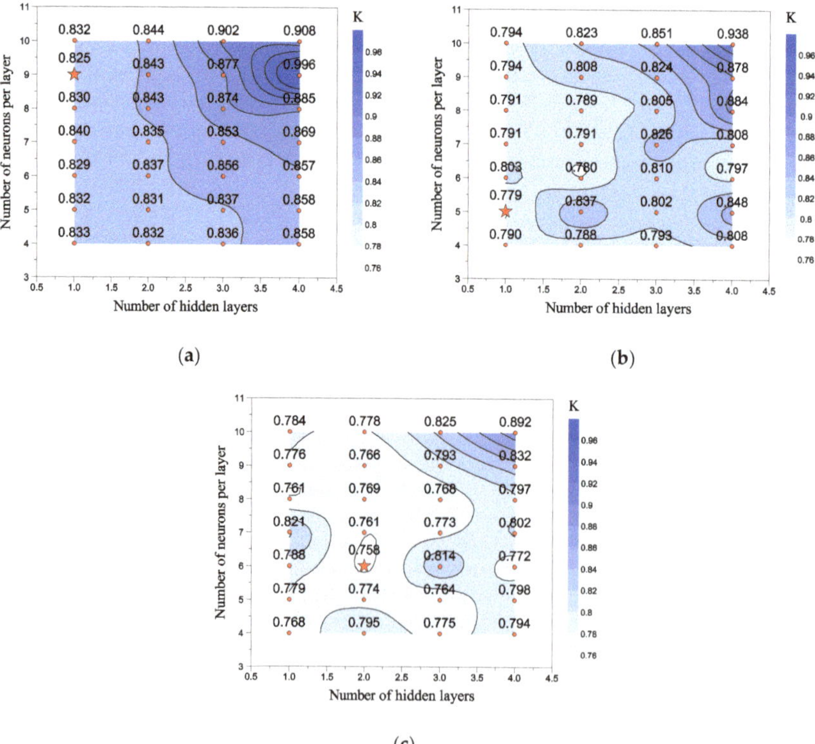

Figure 14. Cost maps for determination of optimal ANN structure: (**a**) ANN_{IC}; (**b**) ANN_{CP}; (**c**) ANN_{IP}.

To intuitively present the prediction performance of ANN models, the predicted CA50 of testing data set are compared with the experimental results as shown in Figure 15. In case that the model can predict CA50 correctly, the data points should be distributed near the black short dotted line. From Figure 15a, when the input parameters are only ion current-related parameters, the prediction accuracy of the ANN_{IC} model is low. The experimental CA50 changes between −15 °CA aTDC and 15 °CA aTDC, but the predicted CA50 is not sensitive enough and only changes between −5 °CA aTDC and 5 °CA aTDC. When the input variables are only cylinder pressure-related parameters, the prediction accuracy improves significantly as more data distribute near the black short dotted line, but there are also several outlier cycles. When the input variables are a combination of ion current and cylinder pressure-related parameters, almost all data points are concentrated, distributing around the black short dotted line, so the prediction accuracy has been further improved.

To get a quantitative comparison result, two evaluation indices are chosen. One is the correlation coefficient between predicted CA50 and experimental CA50, the other is the root mean squared error (RMSE) which has the same dimensions as the output variables. The statistical results are shown in Table 4. The ANN_{IC} model has the worst prediction accuracy as its correlation coefficient is the lowest while the RMSE is the highest. Compared to the ANN_{IC} model, the correlation coefficient of the ANN_{CP} model is enhanced from 0.49 to 0.77, and the RMSE is reduced from 1.40 °CA to 1.06 °CA. For the ANN_{IP} model, it shows the best prediction accuracy as the correlation coefficient further increases from 0.77 to 0.82, at the same time, the RMSE is further reduced from 1.06 °CA to 0.94 °CA.

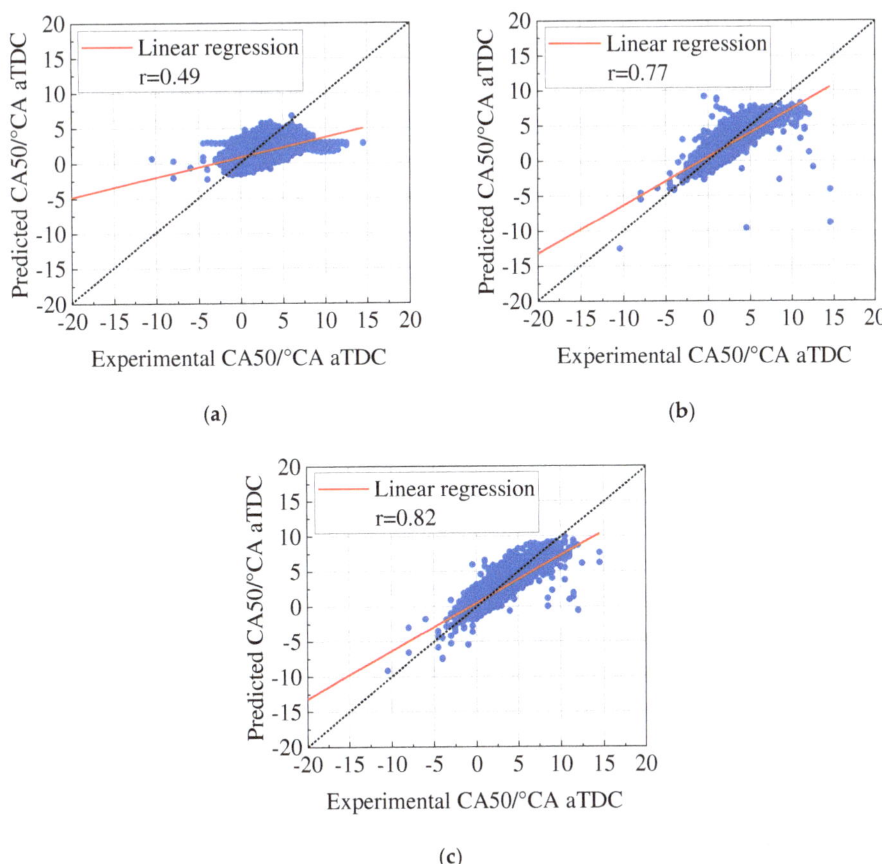

Figure 15. Comparison of predicted CA50 and experimental CA50: (**a**) ANN_{IC}; (**b**) ANN_{CP}; (**c**) ANN_{IP}.

Table 4. Prediction performance of three ANN models.

Evaluation Indexes	ANN_{IC}	ANN_{CP}	ANN_{IP}
r	0.49	0.77	0.82
RMSE/°CA	1.4	1.06	0.94

The above results show that on the one hand, the prediction robustness based on the cylinder pressure is better than based on the ion current signal. A reasonable interpretation is that it is a global measurement quantity that is much less sensitive to the boundary condition. However, on the other hand, the results confirmed that the ion current signal can give significant additional information beyond pressure trace. It should be noted that this is another situation different from the application case in the previous section. In the previous section, only by combining both signals, two kinds of knock can be completely predicted. Meanwhile, in this session, the results show that under normal circumstances, the combination of ion current and cylinder pressures still makes sense. Therefore, it is promising to integrate an ANN-based ion current/cylinder pressure cooperative combustion prediction model to the existing ion current/cylinder pressure cooperative combustion diagnosis and control system developed in this article.

6. Conclusions and Outlook

This article summarizes the following conclusions:

(1) Two kinds of ion current detection systems are comprehensively introduced and compared at the hardware level and signal level. In general, the performance of the DC-power ion current detection system is better than that of the capacitive ion circuit detection system both in SI and HCCI modes, so it is more suitable for laboratory research. The biggest advantage of the capacitive ion current detection system is that it is compatible with mainstream ignition systems, so it is more conducive to industrial applications.

(2) The ion current/cylinder pressure cooperative combustion diagnosis and control system is implemented on the engine prototyping control unit. The accuracy of online calculation results has been validated as its correlation coefficient to the offline calculation results is higher than 0.9. One application case of using this system for HCCI abnormal combustion control and stability improvement under high load boundary condition is introduced. After activating the controller, the standard deviation of CA50 is decreased from 1.88 °CA to 1.75 °CA and the standard deviation of IMEP is decreased from 0.011 MPa to 0.008 MPa. More importantly, the knock intensity of all cycles after water injection control is below the acceptable limit. Hence, with this algorithm, the knock can be effectively suppressed and the combustion stability has been improved.

(3) The potential of ion current/cylinder pressure synergy combined with an artificial neural network (ANN) model for combustion prediction has been evaluated. The ANN_{IC} model (only ion current as input) has the worst prediction accuracy. The prediction accuracy of the ANN_{CP} model (only cylinder pressure as input) is significantly improved, whose correlation coefficient is enhanced from 0.49 to 0.77, and the RMSE is reduced from 1.40 °CA to 1.06 °CA. The ANN_{IP} model (both ion current and cylinder pressure as inputs) shows the best prediction accuracy since the correlation coefficient further increases from 0.77 to 0.82 and the RMSE is further reduced from 1.06 °CA to 0.94 °CA. The results confirmed that the ion current signal can give additional information beyond the pressure trace.

For the next step, the ANN-based ion current/cylinder pressure cooperative combustion prediction model will be integrated into the existing ion current/cylinder pressure cooperative combustion diagnosis and control system developed in this article. There will be more application attempts with this system in SI and HCCI modes in the future.

Author Contributions: Conceptualization, D.Z. and L.L.; formal analysis, D.Z.; funding acquisition, J.D., J.A., and L.L.; investigation, D.Z., H.W., S.W., and H.Z.; methodology, D.Z.; project administration, J.D., J.A., and L.L.; resources, J.D., J.A., and L.L.; software, D.Z., J.W., S.W., and H.Z.; supervision, L.L.; writing—original draft, D.Z.; writing—review & editing, J.A. and L.L. All authors have read and agreed to the published version of the manuscript.

Funding: This research was funded by China Natural Science Foundation (No.52076153 & No.51761135105), German Research Association (the joint Sino-German research project "Ion-Current Sensor based Closed-Loop Control of Lean Gasoline Combustion with High Compression Ratio", No.392430670), and KSPG Professorship Chair of CDHK in Tongji University.

Conflicts of Interest: The authors declare no conflict of interest.

References

1. Reitz, R.D.; Ogawa, H.; Payri, R.; Fansler, T.; Kokjohn, S.; Moriyoshi, Y.; Agarwal, A.K.; Arcoumanis, D.; Assanis, D.; Bae, C.; et al. IJER editorial: The future of the internal combustion engine. *Int. J. Eng. Res.* **2019**, *21*, 1–8. [CrossRef]
2. Kalghatgi, G. Is it really the end of internal combustion engines and petroleum in transport? *Appl. Energy* **2018**, *225*, 965–974. [CrossRef]
3. Chu, S.; Majumdar, A. Opportunities and challenges for a sustainable energy future. *Nature* **2012**, *488*, 294–303. [CrossRef]
4. Luo, H.; Nishida, K.; Uchitomi, S.; Ogata, Y.; Zhang, W.; Fujikawa, T. Microscopic behavior of spray droplets under flat-wall impinging condition. *Fuel* **2018**, *219*, 467–476. [CrossRef]
5. Luo, H.; Nishida, K.; Uchitomi, S.; Ogata, Y.; Zhang, W.; Fujikawa, T. Effect of temperature on fuel adhesion under spray-wall impingement condition. *Fuel* **2018**, *234*, 56–65. [CrossRef]
6. Zhou, F.; Fu, J.; Li, D.; Liu, J.; Lee, C.; Yin, Y. Experimental study on combustion, emissions and thermal balance of high compression ratio engine fueled with liquefied methane gas. *Appl. Therm. Eng.* **2019**, *161*, 114125. [CrossRef]
7. Yang, Z.; Miganakallu, N.; Miller, T.; Worm, J.; Naber, J.; Roth, D. Comparing methods for improving spark-ignited engine efficiency: Over-expansion with multi-link cranktrain and high compression ratio with late intake valve closing. *Appl. Energy* **2020**, *262*, 114560. [CrossRef]
8. Doğan, H.E.; Kutlar, O.A.; Javadzadehkalkhoran, M.; Demirci, A. Investigation of Burn Duration and NO Emission in Lean Mixture with CNG and Gasoline. *Energies* **2019**, *12*, 4432. [CrossRef]
9. Gong, C.; Yi, L.; Zhang, Z.; Sun, J.; Liu, F. Assessment of ultra-lean burn characteristics for a stratified-charge direct-injection spark-ignition methanol engine under different high compression ratios. *Appl. Energy* **2020**, *261*, 114478. [CrossRef]
10. Zhao, L.; Qi, W.; Wang, X.; Su, X. Potentials of EGR and lean mixture for improving fuel consumption and reducing the emissions of high-proportion butanol-gasoline engines at light load. *Fuel* **2020**, *266*, 116959. [CrossRef]
11. Zhen, X.; Li, X.; Wang, Y.; Liu, D.; Tian, Z.; Wang, Y. Effects of the initial flame kernel radius and EGR rate on the performance, combustion and emission of high-compression spark-ignition methanol engine. *Fuel* **2020**, *262*, 116633. [CrossRef]
12. Esfahanian, V.; Salahi, M.M.; Gharehghani, A.; Misrsalim, M. Extending the lean operating range of a premixed charged compression ignition natural gas engine using a pre-chamber. *Energy* **2019**, *119*, 1181–1194. [CrossRef]
13. Tsuboi, S.; Miyokawa, S.; Matsuda, M.; Yokomori, T.; Iida, N. Influence of spark discharge characteristics on ignition and combustion process and the lean operation limit in a spark ignition engine. *Appl. Energy* **2019**, *250*, 617–632. [CrossRef]
14. Pan, M.; Zhu, Y.; Bian, X.; Liang, Y.; Lu, F.; Ban, Z. Theoretical analysis and comparison on supercritical CO_2 based combined cycles for waste heat recovery of engine. *Energy Convers. Manag.* **2020**, *219*, 113049. [CrossRef]
15. Zhu, S.; Zhang, K.; Deng, K. A review of waste heat recovery from the marine engine with highly efficient bottoming power cycles. *Renew. Sustain. Energy Rev.* **2020**, *120*, 113049. [CrossRef]

16. Wick, M.; Bedei, J.; Gordon, D.; Wouters, C.; Lehrheuer, B.; Nuss, E.; Andert, J.; Koch, C. In-cycle control for stabilization of homogeneous charge compression ignition combustion using direct water injection. *Appl. Energy* **2019**, *240*, 1061–1074. [CrossRef]
17. Schmitt, S.; Wick, M.; Wouters, C.; Ruwe, L.; Graf, I.; Andert, J.; Hansen, N.; Pischinger, S.; Höinghaus, K. Effects of water addition on the combustion of iso-octane investigated in laminar flames, low-temperature reactors, and an HCCI engine. *Combust. Flame* **2020**, *212*, 433–447. [CrossRef]
18. Onishi, S.; Jo, S.; Shoda, K.; Jo, P.; Kato, S. *Active Thermo-Atmosphere Combustion (ATAC)—A New Combustion Process for Internal Combustion Engines*; SAE Technical Paper 790501; SAE International: Detroit, MI, USA, 1979.
19. Noguchi, M.; Tanaka, Y.; Tanaka, T.; Takeuchi, Y. *A Study on Gasoline Engine Combustion by Observation of Intermediate Reactive Products during Combustion*; SAE Technical Paper 790840; SAE International: Detroit, MI, USA, 1979.
20. Sauter, J.; Lee, C.; Ra, Y.; Reitz, R. *Model Parameter Sensitivity of Mixing and UHC/CO Emissions in a PPCI, Low-Load Optical Diesel Engine*; SAE Technical Paper 2011-01-0844; SAE International: Detroit, MI, USA, 2011.
21. Reezaei, S.; Zhang, F.; Xu, H.; Ghafourian, A.; Herreros, J.; Shuai, S. Investigation of two-stage split-injection strategies for a Dieseline fuelled PPCI engine. *Fuel* **2013**, *107*, 299–308. [CrossRef]
22. Ansari, E.; Shahbakhti, M.; Naber, J. Optimization of performance and operational cost for a dual mode diesel-natural gas RCCI and diesel combustion engine. *Appl. Energy* **2018**, *231*, 549–561. [CrossRef]
23. Liu, X.; Kokjohn, S.; Li, Y.; Wang, H.; Li, H.; Yao, M. A numerical investigation of the combustion kinetics of reactivity controlled compression ignition (RCCI) combustion in an optical engine. *Fuel* **2019**, *241*, 753–766. [CrossRef]
24. Kawamura, Y.; Shinshi, M.; Sato, H.; Takahashi, N.; Irlyama, M. *MBT Control through Individual Cylinder Pressure Detection*; SAE Technical Paper 881779; SAE International: Detroit, MI, USA, 1988.
25. Draper, C.S.; Li, Y.T. *Principles of Optimalizing Control Systems and an Application to the Internal Combustion Engine*; American Society of Mechanical Engineers: New York, NY, USA, 1951.
26. Fujii, I.; Yagi, S.; Kawai, M.; Yoshikawa, H. *MBT Control Utilizing Crank Angle of Maximum Combustion Pressure*; SAE Technical Paper 890759; SAE International: Detroit, MI, USA, 1989.
27. Zhu, G.; Daniels, C.; Winkelman, J. *MBT Timing Detection and its Closed-Loop Control Using In-Cylinder Pressure Signal*; SAE Technical Paper 2003-01-3266; SAE International: Detroit, MI, USA, 2003.
28. Houpt, P.; Andreadakis, S. *Estimation of Fuel-Air Ratio from Cylinder Pressure in Spark Ignited Engines*; SAE Technical Paper 830418; SAE International: Detroit, MI, USA, 1983.
29. Tunestål, P.; Hedrick, J.K. Cylinder air/fuel ratio estimation using net heat release data. *Control Eng. Pract.* **2003**, *11*, 311–318. [CrossRef]
30. Kumar, M.; Shen, T. In-cylinder pressure-based air-fuel ratio control for lean burn operation mode of SI engines. *Energy* **2017**, *120*, 106–116. [CrossRef]
31. Randall, K.; Powell, J. *A Cylinder Pressure Sensor for Spark Advance Control and Knock Detection*; SAE Technical Paper 790139; SAE International: Detroit, MI, USA, 1979.
32. Sawamoto, K.; Kawamura, Y.; Kita, T.; Matsushita, K. *Individual Cylinder Knock Control by Detecting Cylinder Pressure*; SAE Technical Paper 871911; SAE International: Detroit, MI, USA, 1987.
33. Ravaglioli, V.; Bussi, C. Model-Based Pre-Ignition Diagnostics in a Race Car Application. *Energies* **2019**, *12*, 2277. [CrossRef]
34. Cho, S.; Park, J.; Song, C.; Oh, S.; Lee, S.; Kim, M.; Min, K. Prediction Modeling and Analysis of Knocking Combustion using an Improved 0D RGF Model and Supervised Deep Learning. *Energies* **2019**, *12*, 844. [CrossRef]
35. Shimasaki, Y.; Kobayashi, M.; Sakamoto, H.; Ueno, M.; Hasegawa, M.; Yamaguchi, S. *Study on Engine Management System Using in-Cylinder Pressure Sensor Integrated with Spark Plug*; SAE Technical Paper 2004-01-0519; SAE International: Detroit, MI, USA, 2004.
36. Cesario, N.; Tagliatatela, F.; Lavorgna, M. Methodology for misfire and partial burning diagnosis in SI engines. *IFAC PapersOnline* **2006**, *4*, 1024–1028. [CrossRef]
37. Gürbüz, H. Experimental evaluation of combustion parameters with ion-current sensor integrated to fast response thermocouple in SI engine. *J. Energy Eng.* **2017**, *143*, 04016046. [CrossRef]
38. Zhu, D.; Chen, Z.; Chao, Y.; Deng, J.; Li, L. Primary Study on the Transient EGR Control of GDI Turbocharged Engine by Ion Sensing Technology. *IFAC PapersOnLine* **2018**, *51*, 146–153. [CrossRef]

39. Badawy, T.; Rai, N.; Singh, J.; Bryzik, W.; Henein, N. Effect of design and operating parameters on the ion current in a single-cylinder diesel engine. *Int. J. Eng. Res.* **2011**, *12*, 601–616. [CrossRef]
40. Gao, Z.; Liu, B.; Gao, H.; Meng, X.; Tang, C.; Wu, X.; Huang, Z. The correlation between the cylinder pressure and the ion current fitted with a Gaussian algorithm for a spark ignition engine fuelled with naturalgas–hydrogen blends. *Proc. Inst. Mech. Eng. D J. Automob. Eng.* **2014**, *228*, 1480–1490. [CrossRef]
41. Strandh, P.; Christensen, M.; Bengtsson, J.; Johansson, R.; Vressner, A.; Tunestål, P.; Johansson, B. *Ion Current Sensing for HCCI Combustion Feedback*; SAE Technical Paper 2003-01-3216; SAE International: Detroit, MI, USA, 2003.
42. Xie, H.; Wu, Z.; Sun, Y. Ion current characteristics of gasoline HCCI combustion process based on internal EGR. *J. Tianjin Univ.* **2008**, *41*, 547–552.
43. Dong, G.; Li, L.; Zhu, D.; Deng, J. Ion Current Features of HCCI Combustion in a GDI Engine. *Automot. Innov.* **2019**, *2*, 305–313. [CrossRef]
44. Hellring, M.; Munther, T.; Rögnvaldsson, T.; Wickström, N.; Carlsson, C.; Larsson, M.; Nytomt, J. *Spark Advance Control Using the Ion Current and Neural Soft Sensors*; SAE Technical Paper 1999-01-1162; SAE International: Detroit, MI, USA, 1999.
45. Reinmann, R.; Saitzkoff, A.; Mauss, F. *Local Air-Fuel Ratio Measurements Using the Spark Plug as an Ionization Sensor*; SAE Technical Paper 970856; SAE International: Detroit, MI, USA, 1997.
46. Auzins, J.; Johansson, H.; Nytomt, J. *Ion-Gap Sense in Misfire Detection, Knock and Engine Control*; SAE Technical Paper 950004; SAE International: Detroit, MI, USA, 1995.
47. Fan, Q.; Bian, J.; Lu, H.; Tong, S.; Li, L. Misfire detection and re-ignition control by ion current signal feedback during cold start in two-stage direct-injection engines. *Int. J. Eng. Res.* **2014**, *15*, 37–47. [CrossRef]
48. Chao, Y.; Chen, X.; Deng, J.; Hu, Z.; Wu, Z.; Li, L. Additional injection timing effects on first cycle during gasoline engine cold start based on ion current detection system. *Appl. Energy* **2018**, *221*, 55–66. [CrossRef]
49. Liu, Y.; Deng, J.; Hu, Z.; Li, L. In-cycle combustion feedback control for abnormal combustion based on digital ion current signal. *Int. J. Eng. Res.* **2018**, *19*, 241–249. [CrossRef]
50. Collings, N.; Dinsdale, S.; Eade, D. *Knock Detection by Means of the Spark Plug*; SAE Technical Paper 860635; SAE International: Detroit, MI, USA, 1986.
51. Laganá, A.; Lima, L.; Justo, J.; Arruda, B.; Santos, M. Identification of combustion and detonation in spark ignition engines using ion current signal. *Fuel* **2018**, *227*, 469–477. [CrossRef]
52. Tong, S.; Li, H.; Yang, Z.; Deng, J.; Hu, Z.; Li, L. *Cycle Resolved Combustion and Pre-Ignition Diagnostic Employing Ion Current in a PFI Boosted SI Engine*; SAE Technical Paper 2015-01-0881; SAE International: Detroit, MI, USA, 2015.
53. Wang, J.; Hu, Z.; Zhu, D.; Ding, W.; Li, L.; Yan, W.; Jian, T.; Chen, L. *In Cycle Pre-Ignition Diagnosis and Super-Knock Suppression by Employing Ion Current in a GDI Boosted Engine*; SAE Technical Paper 2020-01-1148; SAE International: Detroit, MI, USA, 2020.
54. Flierl, R.; Gollasch, D.; Knecht, A.; Hannibal, W. *Improvements to a Four Cylinder Gasoline Engine through the Fully Variable Valve Lift and Timing System UniValve®*; SAE Technical Paper 2006-01-0223; SAE International: Detroit, MI, USA, 2006.
55. Wick, M.; Bedei, J.; Andert, J.; Lehrheuer, B.; Pischinger, S.; Nuss, E. Dynamic measurement of HCCI combustion with self-learning of experimental space limitations. *Appl. Energy* **2020**, *262*, 114364. [CrossRef]

Article

Individual Cylinder Combustion Optimization to Improve Performance and Fuel Consumption of a Small Turbocharged SI Engine

Luca Marchitto *, **Cinzia Tornatore** and **Luigi Teodosio**

Istituto di Scienze e Tecnologie per l'Energia e la Mobilità Sostenibili-CNR-via Marconi, 4-80125 Napoli, Italy; cinzia.tornatore@cnr.it (C.T.); luigi.teodosio@cnr.it (L.T.)

* Correspondence: luca.marchitto@cnr.it

Received: 1 October 2020; Accepted: 21 October 2020; Published: 23 October 2020

Abstract: Stringent exhaust emission and fuel consumption regulations impose the need for new solutions for further development of internal combustion engines. With this in mind, a refined control of the combustion process in each cylinder can represent a useful and affordable way to limit cycle-to-cycle and cylinder-to-cylinder variation reducing CO_2 emission. In this paper, a twin-cylinder turbocharged Port Fuel Injection–Spark Ignition engine is experimentally and numerically characterized under different operating conditions in order to investigate the influence of cycle-to-cycle variation and cylinder-to-cylinder variability on the combustion and performance. Significant differences in the combustion behavior between cylinders were found, mainly due to a non-uniform effective in-cylinder air/fuel (A/F) ratio. For each cylinder, the coefficients of variation (CoVs) of selected combustion parameters are used to quantify the cyclic dispersion. Experimental-derived CoV correlations representative of the engine behavior are developed, validated against the measurements in various speed/load points and then coupled to an advanced 1D model of the whole engine. The latter is employed to reproduce the experimental findings, taking into account the effects of cycle-to-cycle variation. Once validated, the whole model is applied to optimize single cylinder operation, mainly acting on the spark timing and fuel injection, with the aim to reduce the specific fuel consumption and cyclic dispersion.

Keywords: combustion optimization; cylinder-to-cylinder variation; cycle-to-cycle variation; fuel consumption; 0D-1D engine modeling; experiments

1. Introduction

Stringent exhaust emission and fuel consumption regulations always impose the need for new solutions for further development of internal combustion engines. Concerning compression ignition engines, widely used due to their high fuel efficiency and specific power output, significant attention is paid to biodiesel, with the aim of both replacing fossil fuels and reducing exhaust emissions [1,2]. On the other hand, the high Nitrogen Oxides and Particulate Matter production represents a limit to meet worldwide exhaust emissions regulation, arousing major interest in spark ignition (SI) engines. The combination of new technologies and solutions such as downsizing/turbocharging, variable valve timing and actuation, cooled exhaust gas recirculation and water injection has allowed a strong optimization of SI engines throughout the recent decades [3–7].

A turbocharged/downsized SI engine can operate close to minimum brake specific fuel consumption (BSFC) over a wide range of speeds and loads. In more detail, this kind of engine provides major reductions in BSFC if compared to a naturally aspirated, port fuel injection counterpart at low-to-moderate torque levels. The consumption reductions come from a drop in friction losses (thanks to the smaller number and dimension of cylinders and related moving mechanisms) and pumping losses (associated with the reduction or abolition of throttling at low loads).

The application of turbocharging and downsizing in commercial vehicles has been increasing since 2000s, but at the present time there are still technical issues connected to this kind of engine such as pre-ignition, knocking tendency and high temperatures at the exhaust. The elevated intake pressures increase the knock predisposition; due to pre-ignition and knock concerns, the spark timing range for acceptable operation is significantly reduced at high Brake Mean Effective Pressures, which results in an increase in fuel consumption. On the other hand, turbochargers are characterized by maximum allowed values of temperature to avoid an excessive thermal stress for the turbine blades; in order to maintain exhaust temperatures below this limit and to reduce the knock tendency at high load and low speed, fuel enrichment is often used. Moreover, the latter control strategy obviously deteriorates the fuel consumption of the vehicle.

In addition to this, another phenomenon that usually becomes a limiting factor for the operational range of the engine design [8] is the cycle-to-cycle variation (CCV). Cycle-to-cycle variation consists of the differences between consecutive combustion events and it is mainly due to the stochastic nature of internal turbulent flow structures and combustion processes [9]; it can lead to a reduction in the engine efficiency because of the potentially associated knocking phenomena, and also increase the production of unburned hydrocarbon and carbon monoxide due to incomplete combustion. Moreover, this phenomenon can induce unwanted fluctuations of power [10].

Due to all these restrictions and limiting factors, the modern turbocharged SI engine needs a robust control through engine control unit (ECU). An ECU control scheme manages engine emissions and fuel economy gathering the data obtained by different sensors and regulating the input parameters such as spark timing, fuel injection timing, etc. Engine control procedures are carried out taking into account the overall engine operation and emissions assuming that the different cylinders have the same behavior. Anyway, although many efforts have been made to introduce in all cylinders a homogeneous air-fuel mixture, some non-uniformities remain that induce the phenomenon of cylinder-to-cylinder variation. This occurrence is attributed to the uneven distribution of the charge in different cylinders caused by thermo-fluid dynamic processes taking place during the intake process. This means that some cylinders work in lean condition and the engine parameters are adjusted to make these cylinders run reliably. On the other hand, the richer cylinders probably produce unburned hydrocarbons and waste fuel. Moreover, this difference in air/fuel (A/F) ratio of each cylinder leads to a further increase in cycle-to-cycle variation as an optimal combustion phasing for all cylinders is impossible without a single cylinder control strategy of engine parameters. A significant cylinder-to-cylinder variability compromises the engine balancing, increasing both longitudinal and torsional vibrations which strain the crankshaft with fatigue stress and decrease comfort and safety of the vehicle passenger [11]. Saxén et al. [12] investigated the effect of cylinder power balancing on torque propagation. They found a reduction of 82% in the standard deviation of the power unbalances at the flywheel through a dynamic control of individual cylinder fuel injection duration and timing. As a consequence, a reduction in the magnitude of all the torque order frequency components below ignition frequency was found.

According to this scenario, a refined control of combustion process in each cylinder can represent a useful and affordable way to limit cylinder-to-cylinder and cycle-to-cycle variations and this would be a significant step towards further optimization of SI engines.

A number of studies are available in the current literature discussing the origin, the main effects and the control of both cylinder-to-cylinder variation and cyclic dispersion for internal combustion engines. Special attention is here devoted to the analysis of literature works discussing the impact of the above phenomena on the operation of Spark-ignition engines.

As reported by Zhou et al. [13], the research works in the field of cylinder non-uniformity in a multi-cylinder engine can be summarized in three main aspects.

The first one is the study of variation of several engine parameters among different cylinders and their influence factors. In particular, indicated mean effective pressure (IMEP) and its coefficient of variation (COV_{IMEP}) were extensively studied as main parameters that can describe the cylinder performance and variations [14].

The second aspect of the literature study is how the cylinder-to-cylinder and the cycle-to-cycle variations influence the performance and emissions of a multi-cylinder engine [15–17].

The last aspect of the research work in this field is how to reduce the above discussed variability in a multi-cylinder engine. The results obtained show that this phenomenon is very difficult to control; some good results were obtained through controlling the inlet air temperature [18], or through the development of a cylinder balancing control strategy [19,20].

An effective way to reduce cycle-to-cycle variations is adding hydrogen to the primary fuel. This solution is considered especially for spark ignition engines fueled with Compressed Natural Gas (CNG) and Liquefied Petroleum Gas (LPG). Hydrogen has a high-octane number and burn velocity, promoting a more stable combustion and reducing knock tendency [21]. These characteristics make the use of hydrogen attractive especially for applications on SI engines working under lean burn conditions as they allow the extension of the lean limit [22].

Of course, certain benefits can be also achieved in terms of reduced CCV and improved combustion stability through the optimal control of cylinder operation, acting on the mitigation or the substantial elimination of the cylinder unbalance in a multi-cylinder SI engine [20–24].

Most of the reference works on this topic are oriented towards an experimental approach. Fewer numerical analyses can be found in the literature, even if the use of numerical methodologies is receiving a growing interest on the study of individual cylinder operation and variability to reduce the time and cost required by the experiments. The combination of numerical and experimental methods appears to be a desirable choice to properly forecast the individual cylinder behavior and, subsequently, to explore the engine control strategies capable of improving the cylinder non uniformities and the related cyclic dispersion. Furthermore, the selection of the numerical approach holds an important role in satisfactorily forecasting the in-cylinder processes without burdening the computational time. From this point of view, among the existing numerical models, a 1D approach offers the possibility to study the individual cylinder behavior in a multi-cylinder engine, showing good balance between accuracy and computational effort. A 1D model is suitable for performing system analyses, testing modified engine architectures, and exploring and optimizing the operation of each cylinder in various engine conditions. On the other hand, a 1D model cannot provide information about local phenomena occurring in the combustion chamber like mixture inhomogeneity, variations in turbulence etc.

This paper presents experimental and 1D numerical investigations performed on a twin-cylinder turbocharged Port Fuel Injection (PFI) Spark Ignition engine in order to study the influence of cycle-to-cycle and cylinder-to-cylinder variations on combustion and performance under different operating conditions. In addition, the developed numerical model is also utilized in a predictive way to make uniform the operation between cylinders in order to optimize engine performance, combustion and CCV.

As is well known, the limits of the 1D model concerning local phenomena predictions do not allow the identification of variations in turbulence and mixture quality near the spark plug even if this has a significant influence on CoV and CCV.

The innovative contribution of this work with respect to the state-of-art papers mainly consists of the prediction of cylinder-related and cycle-related variabilities by means of a 1D model integrated with refined 0D in-cylinder sub-models. A further relevant aspect of the proposed study consists of the integration of the developed engine model with empirical CCV correlations to furnish a first attempt estimation of improved CCV levels resulting from the optimization of engine operation. Following the above discussed numerical procedure, in this paper the experimental tests on the considered SI engine are first presented.

For each cylinder, the coefficients of variations (CoVs) of selected pressure parameters were used to quantify the cyclic dispersion. Experimentally-derived correlations, representative of the actual engine system, were coupled to an advanced 1D model of the whole engine, developed within a commercial code. Once validated, the model was applied to optimize the single cylinder operation, mainly acting on the spark timing and fuel injection, with the aim to reduce the indicated specific fuel consumption (ISFC) and to improve the combustion stability through the optimization of the cyclic dispersion level.

2. Engine Features, Experimental Activity and Procedure

The SI internal combustion engine used in this work is sketched in Figure 1 and its main features, including performance, geometrical and intake/exhaust valve characteristics, are listed in Table 1. It consists of two cylinders coupled to a small waste-gated turbocharger, which allows it to match the boost level for the prescribed full load performance target. Port fuel injectors are mounted along the intake runners just upstream of the valves to feed the engine with liquid gasoline. The engine is also equipped with a Variable Valve Actuation (VVA) device for a flexible control of the intake lift strategy while the exhaust valves present both fixed lift and timing.

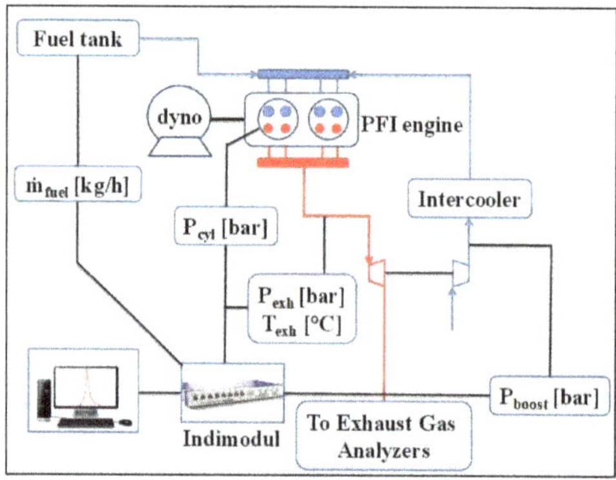

Figure 1. Experimental test bench of the considered spark ignition engine.

Table 1. Engine main features.

2-Cylinder PFI Turbocharged Spark Ignition Engine	
Bore	80.5 mm
Stroke	86 mm
Displacement	875.4 cm^3
Compression ratio	10.0
Fuel	Gasoline, RON 95
Valve number	2 intake/2 exhaust valves per cylinder
EVO/EVC @2 mm lift	134/382 CAD AFTDC
IVO/IVC @2 mm lift	342/356–420/624 CAD AFTDC
Max Brake Torque	146.7 Nm @ 2000 rpm
Max Brake Power	63.7 kW @ 5500 rpm

The engine was installed at a test bench and instrumented with various sensors/transducers. Two piezo-quartz pressure transducers (measuring range 0: 250 bar, accuracy of ±0.1%) in the cylinders allowed the measuring of instantaneous pressure. For combustion analysis, signals were recorded over 270 consecutive cycles, with a resolution of 0.1 Crank Angle Degree (CAD) within the angular window between −90 and 90 CAD After Top Dead Center (AFTDC), assuming a polytrophic thermodynamic process. A 1.0 CAD sampling resolution was set outside this angular interval.

Intake, boost and turbine inlet pressure were monitored through piezo-resistive low-pressure indicating sensors (measuring range 0: 10 bar, accuracy of ±0.1%); intake and exhaust temperatures were measured with thermocouples. The relative A/F ratio (λ) is measured with a sensor located at the engine exhaust, downstream of the turbine, with an accuracy of ±0.01 at λ = 0.8, ±0.007 at λ = 1.0.

The experimental activity can be divided into two phases: spark sweep analysis and ECU conditions analysis.

2.1. Phase 1: Spark Sweep Analyses

The first part of the experimental activity includes spark timing sweeps performed at various loads, a wide range of engine speeds and also stoichiometric and rich A/F mixture conditions at medium/high loads. In particular, a prevailing number of medium/high load conditions is considered in these studies and only a more limited set of part-load cases is analyzed. Table 2 reports the overall test matrix for this phase of experiments consisting of 50 operating points. For each engine speed and selected λ, a spark timing sweep is actuated taking into account a proper range of values. This range includes a reference spark advance corresponding to the Maximum Brake Torque (MBT) at low load or to the heuristic-based knock limit at high load. The richer A/F ratio is investigated at high load regardless of the engine speed in order to take into account the characteristic limitations of the examined engine related to the avoidance of knock occurrence and to the constraint on the maximum allowable temperature at turbine inlet (below 950 °C). It is the case to also underline that the relative A/F ratio is maintained at a constant level for each combination of engine speed, spark timing and IMEP reported in Table 2.

Table 2. Test matrix of spark timing sweep at various speeds, indicated mean effective pressure (IMEP) levels and relative air/fuel ratio (λ).

Speed, rpm	Load	λ, -	SA, CAD AFTDC	IMEP, bar
2160	High	1.00	−7.5/−5.5/−2.5/−1.5/−0.5	15.2/14.8/14.5/14.1/13.7
2160	High	0.96	−6.5/−5.5/−3.5/−2.5/−1.5/−0.5	15.4/15.1/14.9/14.6/14.2/13.9
2500	Low	1.00	−42/−38/−34/−30/−26	4.0/4.1/4.2/4.2/4.2
3000	High	1.00	−14/−13/−12	16.8/16.6/17.1
3000	Low	1.00	−40/−36/−32/−28/−24	4.3/4.5/4.6/4.5/4.5
3500	High	0.92	−16/−14/−12/−10/−8	14.1/13.9/13.7/13.4/13.0
3500	High	1.00	−16/−14/−12/−10/−8	13.9/13.7/13.5/13.1/12.7
4000	High	0.92	−12.5/−11.5/−9.5	15.0/14.9/14.5
4000	Low	1.00	−36.5/−32.5/−28.5/−24.5/−20.5	5.5/5.6/5.6/5.7/5.5
4500	High	0.96	−20.5/−19.5/−18.5/−16.5	13.1/13.1/13.0/12.9
4500	High	1.00	−20.5/−19.5/−18.5/−16.5	12.9/12.9/12.8/12.7

For each analyzed point belonging to the spark sweep analyses, 270 consecutive in-cylinder pressure traces were measured for both engine cylinders, the recorded sequence is then post-processed to obtain the distributions of both pressure variables (i.e., IMEP and p_{max}) and combustion parameters (i.e., mass fraction burned at relevant crank angles, MFB_{10}, MFB_{50} and MFB_{90}) for the individual cylinders of the engine. A standard deviation below 2% was measured for the IMEP of the engine in all the tested operating points. Even lower values were observed in the low/medium load range. The in-cylinder pressure peaks (p_{max}) of two cylinders have a quite similar standard deviation at varying the investigated speed/load points. Standard deviations in the range 4.8–9.9% (1.7–6.5 bar absolute value) and 3.6–9.4% (1.4–6.6 bar absolute value) were recorded for Cyl #1 and Cyl #2, respectively. Similar absolute standard deviations were found for the combustion phasing MFB_{50}: 1.5–2.6 CAD for Cyl #1 and 1.5–2.4 CAD for Cyl#2. Concerning the core combustion durations ($MFB_{10–50}$), a reduced standard deviation was detected with a lower variability range for Cyl #2 (0.73–1.15 CAD) compared to Cyl #1 (0.84–1.51 CAD). The air flow rate was derived by the intake plenum pressure, measured through a piezo-resistive low pressure sensor (measuring range 0: 10 bar, accuracy of ±0.1%), showing a standard deviation of about 3% in almost the whole test matrix.

2.2. Phase 2: Experimental Points on the Engine Map

In a second stage, the ECU standard calibration is considered: the experiments were carried out under the conditions reported in Table 3. The test grid consists of 15 operating points, arranged in three groups characterized by the same speed, null external Exhaust Gas Recirculation (e-EGR) and increasing IMEPs. The matrix in Table 3 was developed considering the engine points more representative of everyday driving conditions. First of all, it was considered that the tested engine, coupled to a segment A vehicle, works along a WLTC driving cycle in a region of the operating plan characterized by a speed from 1000 to 3000 rpm and loads up to 15 bar Brake Mean Effective Pressure (BMEP). For this reason, the speeds of 2000 and 3000 rpm were chosen with an IMEP sweep from 5 to 13 bar and from 5 to 18 bar, respectively.

Table 3. Test matrix of operating points in the engine map.

Group	Label	#	Speed, rpm	IMEP, bar	λ, -	e-EGR, %	SA, CAD AFTDC
Low	2000@5	1	2000	5.0	1.00	0.0	−35.0
	2000@7	2	2000	7.1	1.00	0.0	−29.0
	2000@9	3	2000	9.1	1.00	0.0	−25.0
	2000@11	4	2000	11.0	1.00	0.0	−16.0
	2000@13	5	2000	13.0	1.00	0.0	−12.0
Medium	3000@5	6	3000	5.2	1.00	0.0	−33.0
	3000@7	7	3000	7.0	1.00	0.0	−30.0
	3000@12	8	3000	12.2	0.96	0.0	−11.0
	3000@16	9	3000	16.3	0.90	0.0	−9.0
	3000@18	10	3000	17.8	0.90	0.0	−11.0
High	4000@7	11	4000	7.1	1.00	0.0	−35.0
	4000@9	12	4000	9.1	1.00	0.0	−25.0
	4000@11	13	4000	11.2	1.00	0.0	−21.0
	4000@13	14	4000	13.1	1.00	0.0	−22.0
	4000@16	15	4000	16.1	0.89	0.0	−12.0

Moreover, further operating conditions, outside of the WLTC, typical of highway driving conditions were added: an engine speed of 4000 rpm was selected and the IMEP was changed in a discrete way from 7 to 16 bar.

For each point, main performance and calibration variables are reported and a label is also used to synthesize the speed/load data. An analysis of the investigated ECU operating points highlights a mixture over-fueling at higher IMEP only for medium/high speeds. Furthermore, the spark timing is gradually delayed at increasing load. These control strategies are mandatory for the considered downsized engine to avoid knocking and to limit the thermal stress at turbine inlet. However, the discussed control strategies at high loads greatly penalize engine fuel consumption as emerged from the experimental results presented in the subsequent section of model validation. In the latter section, the experimental outcomes for both cylinders in each operating point of the test matrix (Table 3) show a systematic cylinder-to-cylinder variation that can be easily observed by the differences in the pressure (peak and instantaneous trace) and combustion data of two cylinders. This systematic cylinder-to-cylinder variability is mainly ascribed to the different mass of fuel injected between cylinders. In fact, a comparison between pressure traces from Cyl #1 and Cyl #2 highlights an overlap in the compression stage, indicating an equal air volumetric efficiency for the two cylinders, while Cyl #2 provides higher pressure peak and combustion rate than Cyl #1, suggesting a difference in fuel supply to cylinders. These experimental outcomes suggest that engine cylinders operate under different A/F ratios: lean mixture is realized for Cyl #1 and rich mixture is obtained for Cyl #2 compared to the overall engine A/F ratio.

3. Modeling Approach and Validation

Based on the results of the previously discussed experimental activities, a modeling approach is employed here consisting of the integration of a complete 1D fluid-dynamic model of the whole engine and of properly developed correlations capable of quantifying the cycle-to-cycle variation of Indicated mean effective pressure (IMEP) and In-cylinder pressure peak (p_{max}). Although the adopted 1D modeling can be considered a simple numerical method, it shows, on one hand, the advantage of a reduced computational effort and, on the other hand, it is enhanced with advanced 0D sub-models to refine the description of in-cylinder processes also including the cyclic dispersion (CCV). The CCV sub-model is capable of reproducing on a physical basis the measured sequence of consecutive pressure traces in each cylinder. In a first step, 1D engine model was validated against the experimental data considering both the overall performance of the engine and the ensemble-average in-cylinder pressure cycles. The CoV correlations of IMEP and p_{max} with average combustion parameters were developed by numerical fitting of the outcomes deriving from the post-processing of experimental spark-sweep tests. Then the correlations were validated with the measured data set already adopted for the 1D model validation (Table 3). Once validated, the combination of the 1D engine model and of the experimental CoV correlations allows the provision of a combustion optimization of the individual engine cylinder. In addition, this numerical methodology provides benefits in terms of engine fuel economy and cyclic dispersion level, as well. A detailed analysis of the employed numerical models, of their potentialities and limitations, are reported in the following sub-sections.

3.1. D Engine Modeling

The geometry of the analyzed engine is fully schematized in GT-Power 1D commercial code, through the combination of 0D and 1D modeling approaches. In particular, the intake and exhaust sub-systems described through a 1D approach to properly reproduce the flow and the pressure waves propagation. The typical flow equations (Navier–Stokes) are solved in each pipe of intake and exhaust manifolds. This means that the solutions of continuity, momentum, enthalpy and energy conservation equations are provided. The conservation equations solved by GT-Power are shown below:

Continuity:

$$\frac{dm}{dt} = \sum\nolimits_{boundaries} \dot{m}, \tag{1}$$

Energy:

$$\frac{d(me)}{dt} = -p\frac{dV}{dt} + \sum\nolimits_{boundaries} \left(\dot{m}H\right) - hA_s\left(T_{fluid} - T_{wall}\right), \tag{2}$$

Enthalpy:

$$\frac{d(\rho HV)}{dt} = \sum\nolimits_{boundaries} \left(\dot{m}H\right) + V\frac{dp}{dt} - hA_s\left(T_{fluid} - T_{wall}\right), \tag{3}$$

Momentum:

$$\frac{d\dot{m}}{dt} = \frac{Adp + \sum_{boundaries} \left(\dot{m}u\right) - 4C_f\frac{\rho u|u|Adx}{2D} - K_pA\left(\frac{1}{2}\rho u|u|\right)}{dx}, \tag{4}$$

where $\dot{m}$, m, V, p, ρ, e, H and u are mass flux, mass, volume, pressure, density, specific internal energy, enthalpy and velocity at boundary, respectively. A and A_s are the flow area and the heat transfer surface area, T_{fluid} and T_{wall} the fluid and wall temperatures, h the heat transfer coefficient and D the equivalent diameter. C_f and K_p are the friction and pressure loss coefficients, while dx and dp represent the discretization length and the pressure differential across dx.

The in-cylinder processes, including fuel evaporation, air/fuel mixture formation, turbulent combustion, knock and heat transfer are reproduced based on a 0D modeling. Concerning the port fuel injection, it is actuated by a typical "injector object" imposing the air-fuel ratio.

A value equal to 30% of the injected gasoline is assumed to instantaneously evaporate after the injection event, as results from the advised setting of the adopted 1D code. Furthermore, PFI injection is modeled by neglecting the liquid wall film generation along the intake pipes and the spray dynamics. The cylinder head flow permeability is described through the definition of the steady flow coefficients measured at the engine test bench, both in forward and reverse flow conditions. Turbocharger operation is characterized by means of the measured performance maps of compressor and turbine components. The mechanical friction losses of the engine are reproduced by an empirical correlation according to the Chen-Flynn model. Friction correlation includes the variability of the engine speed and on the in-cylinder peak pressure. This correlation is validated through the available experimental data, demonstrating a high accuracy level. The following friction correlation for FMEP (friction mean effective pressure) is employed here:

$$FMEP = 0.876 + 2.66 \times 10^{-3} \cdot p_{max,cyl} - 6.9 \times 10^{-2} \cdot v_{pm} + 8.1 \times 10^{-3} \cdot v_{pm}^2, \tag{5}$$

where $p_{max,cyl}$ is the cylinder peak pressure and v_{pm} the mean piston speed.

As detailed above, the in-cylinder phenomena are simulated by advanced sub-models, implemented into the code under user routines. The combustion event is reproduced by a 0D two zone (burned and unburned gas) sub-model, where the burning rate term is computed through the well-known 'fractal' approach. According to the fractal model, the burning rate term is written as:

$$\frac{dm_b}{dt} = \rho_u A_T S_L = \rho_u A_L \sum S_L, \tag{6}$$

where ρ_u is the unburned gas density, A_L and A_T the area of the laminar and turbulent flame fronts, respectively, S_L the laminar flame speed (LFS) and Σ the wrinkling factor.

It was demonstrated to present a high level of accuracy in describing the turbulent flame front propagation inside the conventional combustion chamber of spark–ignition engines [25]. As emerged from the fractal theory available in the scientific literature [26], some relevant turbulence parameters are included into the wrinkling factor (Σ) of the fractal-derived burn rate expression. This requires a proper evaluation of the above mentioned in-cylinder turbulence variables by means of a phenomenological turbulence sub-model. To this aim, the combustion model is coupled to a turbulence one. The latter represents a K-k-T model, which solves three balance equations to furnish the engine cycle evolution of mean flow kinetic energy (K), turbulent kinetic energy (k) and tumble vortex momentum (T) [27]. In previous authors' works [27,28], the adopted turbulence model was proved to adequately estimate the in-cylinder turbulence parameters at varying the engine speeds and the valve strategies, without a model tuning variation. A kinetically-derived LFS correlation, included within the fractal burning rate, is adopted because of its better prediction capabilities with respect to the experimentally-derived LFS formulations [29]. LFS correlation is obtained by fitting 1D LFS computations via a chemical kinetic solver and it accounts for the in-cylinder thermodynamic state, equivalence ratio and charge dilution.

Finally, A_L is estimated by an automatic procedure implemented into CAD software processing the actual 3D geometry of the engine combustion chamber.

Referring to the cycle-to-cycle variation model, starting from the input data of CoV of the in-cylinder pressure peak (CoV_{pmax}), representative *faster-than-average* and *slower-than-average* cycles are computed basing on a proper algorithm. This algorithm, extensively discussed in [30], derives *faster* and *slower* burning profiles by the average fractal-predicted burn profile through two stretch factors. The obtained burnt profiles (high, average and low) correspond to three pressure traces. For each in-cylinder pressure cycle, the related maximum value (p_{max}) is computed. The developed CCV model automatically modifies the stretch factors until the target levels of p_{max} for high and low cycles are realized. In this way, the measured CoV_{pmax} imposed as input data in the model is numerically reproduced. It is worth emphasizing that the adopted CCV model was also demonstrated to be very efficient in terms of computational effort [30].

Knock occurrence is detected in the model by the auto-ignition (AI) calculation of the unburned air/fuel mixture. AI is computed by using a tabulated approach [31]. The AI table is obtained by chemical kinetics simulations of auto-ignition, performed in a homogeneous reactor at constant pressure (CP) conditions. In particular, AI table is generated through chemical kinetics simulations carried out by considering the kinetic scheme of Andrae [32]. The AI time is stored in the corresponding table as a value depending on the in-cylinder thermodynamic condition (pressure and temperature), mixture quality (equivalence ratio), and residual content (internal EGR). Knock occurrence is recognized when the AI Integral exceeds a prescribed threshold level [31]. Concerning the heat transfer modeling, the in-cylinder heat transfer (gas-to-wall) is described by the modified Hohenberg correlation [33], while convective, conductive and radiative heat transfer models are considered for the exhaust pipes to improve the prediction of temperature at Turbine Inlet. A standard finite element (FEM) approach, implemented into the 1D code, allows the computation of both piston, cylinder liner and head temperatures by assigning reasonable initial temperature levels. The Hohenberg correlation includes the dependency on pressure and volume of gas and also accounts for the cylinder geometry (bore) and mean piston speed. The overall convection multiplier of the Hohenberg model was selected in order to reach a good experimental/numerical agreement for the in-cylinder pressure cycles at different speed/load points, especially in the expansion phase.

The convective heat transfer coefficients for engine coolants (water and oil) are estimated by proper simulations of coolant circuits and subsequently implemented into the 1D code, assuming a dependency on the engine speed according to an assigned power law [34]. Cooling boundary conditions (temperatures for oil and water) are imposed in the model according to the levels recommended by the engine manufacturer.

The 0D quasi-dimensional combustion model was tuned by a trial-and-error procedure to achieve the best agreement with the in-cylinder pressure cycles. It should be stated that a unique set of tuning constants was selected for all the tested operating conditions.

3.2. Model Validation

The developed 1D model is utilized to simulate the measured operating points on the engine domain, as reported in Table 3. Regarding the setup of the model, the experimental values of combustion phasing (MFB_{50}), air/fuel (A/F) ratio and intake valve strategy (inlet valve closing, IVC) are imposed in the 1D model simulations. In addition, the measured IMEP level is matched by a proportional integral derivative (PID) controller acting on the throttle valve opening at low loads or on the waste-gate valve opening at high loads. EGR valve was not regulated during the simulation and it was fixed to a substantially closed setting since the experimental points to be reproduced were investigated at a null external EGR rate. The model reliability is proved here in terms of overall engine performance variables, combustion characteristics, in-cylinder pressure traces and cycle-to-cycle variation. The numerical/experimental assessments of global performance parameters such as Air flow rate, ISFC, and In-cylinder pressure peak for two cylinders are reported in Figures 2 and 3.

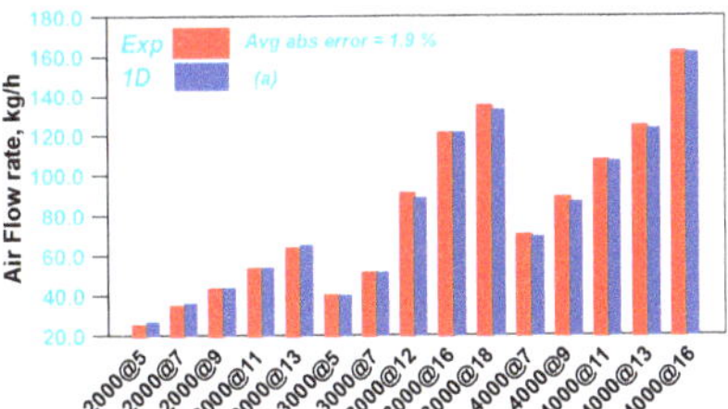

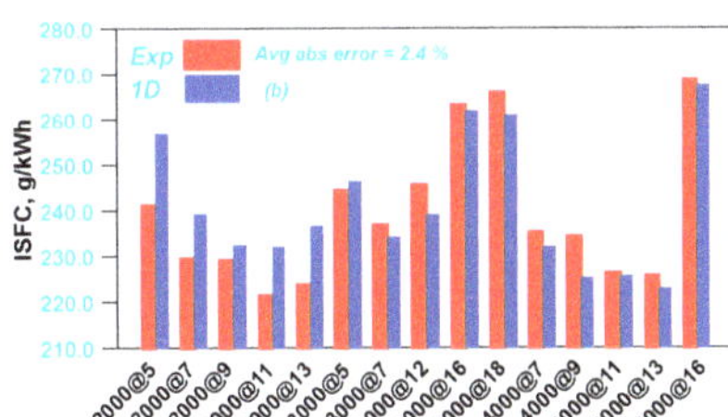

Figure 2. Experimental/Numerical comparison of Air Flow rate (**a**) and indicated specific fuel consumption (ISFC) (**b**) at various engine operating points.

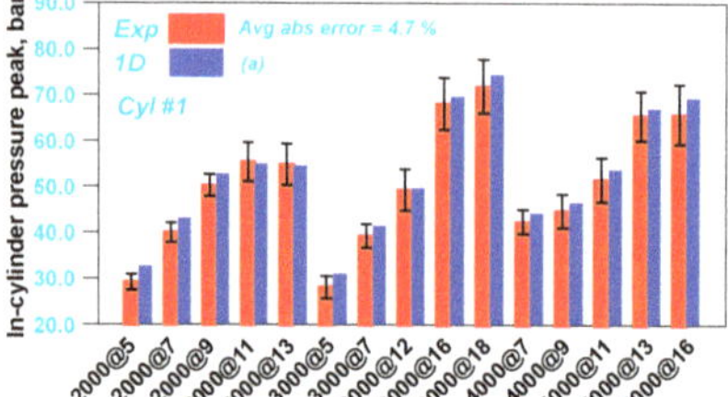

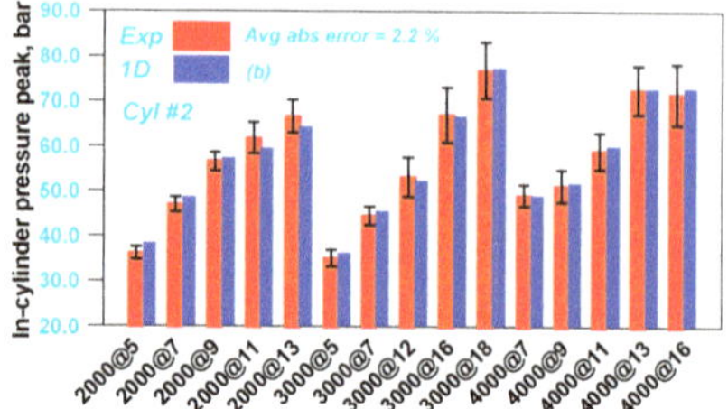

Figure 3. Experimental/Numerical comparison of In-cylinder pressure peak of Cyl#1 (**a**) and Cyl #2 (**b**) at various engine operating points.

The air flow rate (Figure 2a) is predicted with an acceptable accuracy (average absolute percent error of 1.9%), denoting an adequate schematization of the engine geometry. A good experimental/numerical correlation is also realized for the ISFC (average absolute percent error of 2.4% in Figure 2b), hence demonstrating the model capability to properly take into account the combined effects of flow, combustion, and in-cylinder heat transfer phenomena.

The numerical in-cylinder pressure peaks of two cylinders are in a satisfactory agreement with the experimental counterparts. Referring to the in-cylinder pressure peak predictions, larger errors arise for Cyl #1 compared to Cyl #2. However, the average absolute percent error for the Cyl #1 case is below the maximum allowable percent level of 5%, which is considered by the authors to preserve a valuable model accuracy.

The reliability of the combustion modeling is demonstrated by the outcomes reported in Figure 4a,b that shows the experimental/numerical assessments of the combustion core duration (MFB $_{10-50\%}$) for Cyl #1 (Figure 4a) and Cyl #2 (Figure 4b). The combustion sub-model correctly reproduces the combustion development for the selected engine operating conditions (speed/load point, A/F ratio), mainly including the effects of IMEP variation on combustion evolution. It is should be highlighted that a reduced error in the reproduction of the combustion durations has to be considered a fundamental pre-requisite to ensure a proper application of the CCV correlations, as discussed in the next section. The predictivity of the combustion model is also proved by the experimental/numerical assessments of in-cylinder pressure traces for two cylinders shown in Figures 5–7. They refer to three different engine speeds and loads, labelled as 2000@13, 3000@18 and 4000@7, without the activation of the external EGR circuit (i.e., e-EGR% = 0%). In the analyzed operating points, some disagreements arise along the expansion phase of pressure cycles for Cyl #1, probably due to minor inaccuracies in forecasting the combustion tail and/or the heat transfer. However, from a global point of view, numerical in-cylinder pressure cycles present a good agreement with the experimental traces. A similar accuracy is detected for the other investigated speed/load points (Table 3), even if the related plots are not shown here for brevity. In addition, the CCV model is applied to the experimental operating points. To this aim, the experimentally derived coefficient of variation of the in-cylinder peak pressure (CoV_{pmax}) for two cylinders is imposed in the simulations and the computed cycles (high and low ones), representative of the cycle-to-cycle variation, and these are compared to the measured sequence of 270 consecutive pressure cycles. As an example, Figure 8 shows the experimental/numerical assessments for pressure cycles in a medium/low load point and medium speed, i.e., 3000@7. The figure demonstrates that the numerical high and low cycles are in a satisfactory agreement with the extreme *faster-than-average* and *slower-than-average* experimental cycles, respectively. The consistency of the CCV modeling approach is also tested in other speed/load points listed in Table 3. Although not reported here for brevity, in many cases the shape of numerical high and low cycles is close to the measured ones. Of course, inaccuracies in the computation of the ensemble-average in-cylinder pressure trace may reflect CCV prediction, too. Once validated, the CCV model can be used in a predictive way to reproduce the cycle-to-cycle variation of operating points different from those considered in the validation phase. To this aim, the CCV model requires the CoV equation capable of quantifying the variability on the in-cylinder pressure peak. The latter correlation together with the one of IMEP variability will be extensively discussed in terms of generation and validation in the next paragraph.

3.3. Experimentally-Derived Correlations for Cyclic Dispersion and Validation

In the present section, the experimental CCV is extensively analyzed with main aim to define its relationship with the representative engine combustion variables. For this purpose, it is worth underlining that CCV is usually quantified through the Coefficient of Variation (CoV) of the pressure-related parameters, such as the indicated mean effective pressure (IMEP) and the in-cylinder pressure peak (p_{max}). In particular, the first one plays a certain role in the drivability of the engine, while the second one has an influence on the knock occurrence. Based on the available experimental results reported in Table 2, the cycle-to-cycle variation was preliminarily analyzed for each engine cylinder at varying spark timing. As an example, the following Figure 9a,b and Figure 10a,b show the spark timing influence on the dispersions of IMEP and p_{max}, considering operations at low load and at speed of 3000 rpm. In these figures, the average values of IMEP and p_{max} and the related CoVs are reported as continuous lines while the corresponding cycle-derived levels as vertical dots.

As expected, an increasing cycle-to-cycle variation is observed moving the spark timing towards the expansion phase, due to the worsening in the combustion development. For each cylinder, a greater p_{max} variation is realized with respect to the IMEP variation. A comparison between the outcomes of two cylinders (Figures 9 and 10) highlights that Cyl #2 (rich A/F mixture) shows lower CoV levels than Cyl#1 (lean A/F mixture) when varying the spark timing. Figures 9 and 10 also confirm the experimental evidence of higher IMEP and p_{max} values attained by the richer Cyl #2 if compared to those realized by Cyl #1.

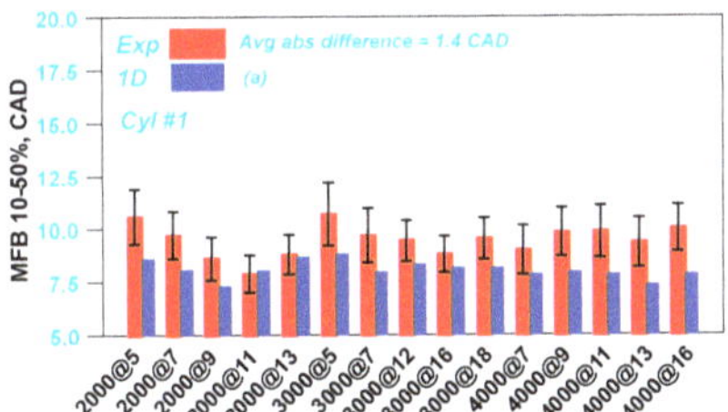

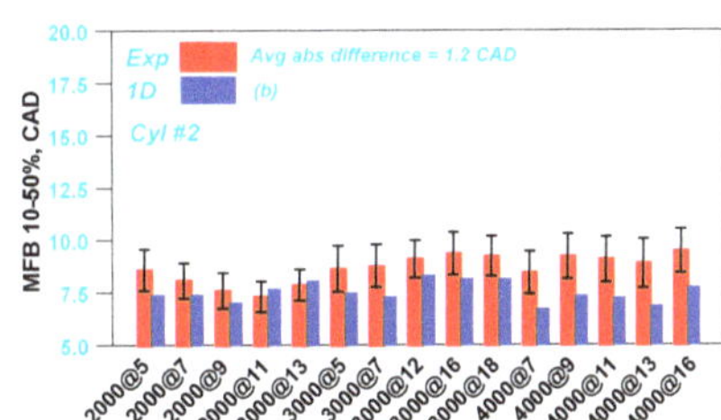

Figure 4. Experimental/Numerical comparison of combustion core duration mass fraction burned (MFB) 10–50% for Cyl#1 (**a**) and Cyl #2 (**b**) at various engine operating points.

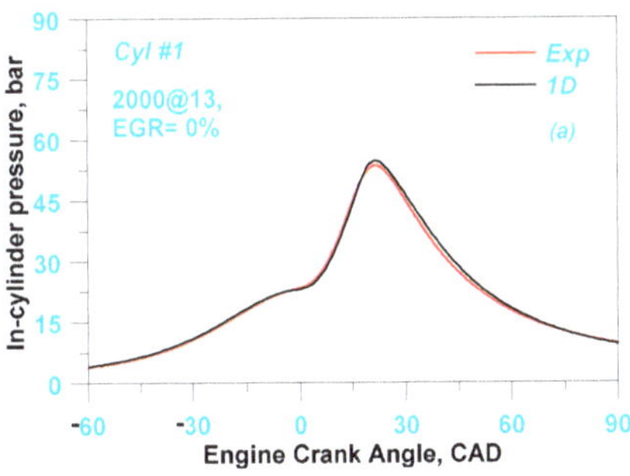

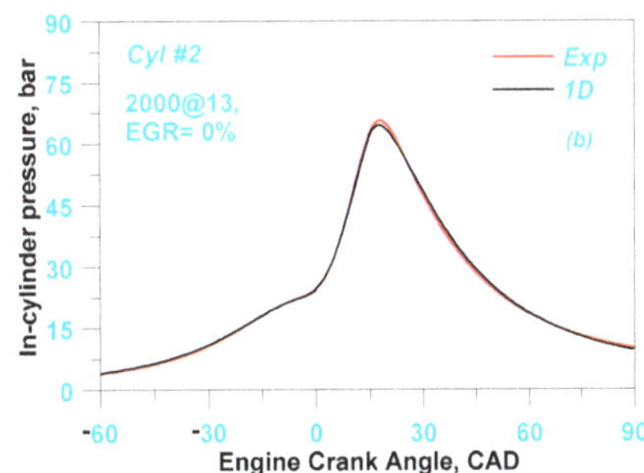

Figure 5. Experimental/Numerical comparison of in-cylinder pressure traces for Cyl#1 (**a**) and Cyl #2 (**b**) at 2000@13.

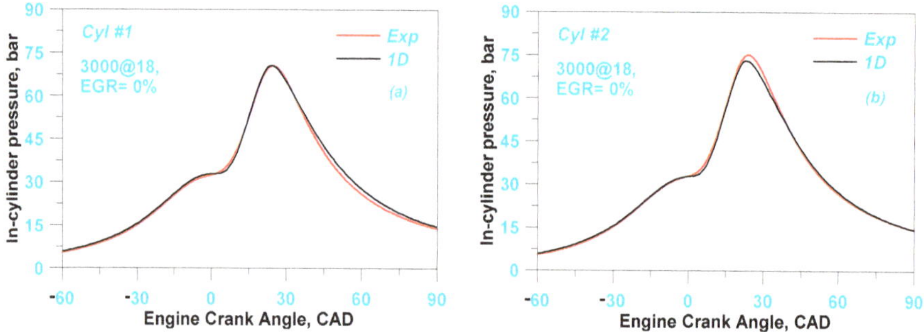

Figure 6. Experimental/Numerical comparison of in-cylinder pressure traces for Cyl#1 (**a**) and Cyl#2 (**b**) at 3000@18.

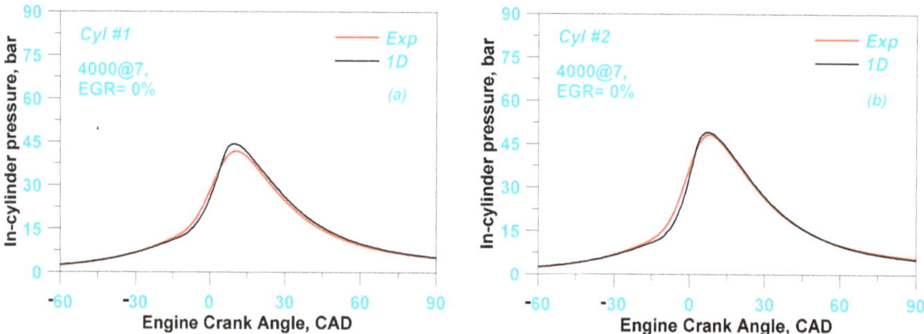

Figure 7. Experimental/Numerical comparison of in-cylinder pressure traces for Cyl#1 (**a**) and Cyl#2 (**b**) at 4000@7.

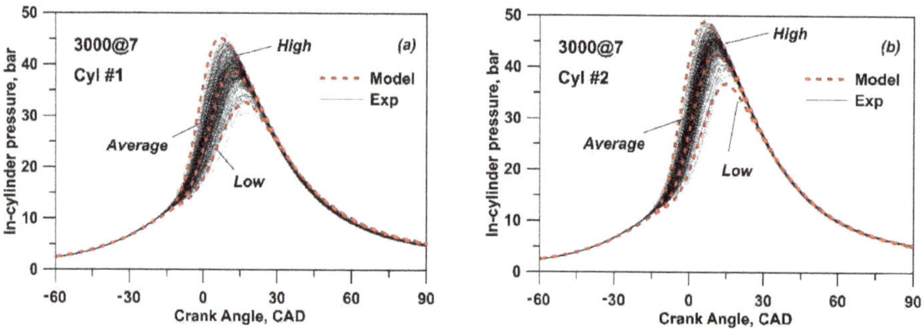

Figure 8. Experimental/Numerical comparison of cycle-to-cycle variation for Cyl#1 (**a**) and Cyl #2 (**b**) at 3000@7.

Similar considerations can be applied to the other analyzed points of the spark timing sweep (Table 2), but the related experimental outcomes are not reported here for sake of brevity. In addition, as already demonstrated in a previous authors' work [35], the CoV trend does not substantially depend on the engine speed. On the other hand, the cycle-to-cycle variation shows a certain dependency on the combustion-related parameters, i.e., combustion phasing and characteristic duration. In order to optimize the engine's overall performance, the optimization of each cylinder operation was realized to eliminate the cylinder-to-cylinder non uniformities. In this perspective, the relationship between CCV characteristics and the combustion-related variables is analyzed here by averaging the experimental data of the two engine cylinders. In this way, experimental-derived CCV correlations can be defined and referred to the examined engine. Once validated, engine CCV correlations can be easily employed to forecast the cyclic dispersion under optimized operating conditions for both cylinders. Based on the above considerations, the following Figures 11 and 12 show the dependency of the engine CoV_{IMEP} and CoV_{pmax} on the averaged combustion parameters, including 50 operating points (Table 2) and the data measured in all cylinders.

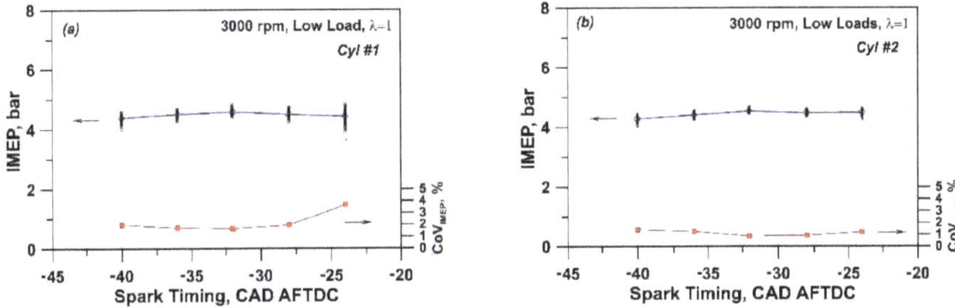

Figure 9. IMEP and coefficient of variance $(CoV)_{IMEP}$ vs. Spark timing for Cyl#1 (**a**) and Cyl#2 (**b**) at 3000 rpm and low load.

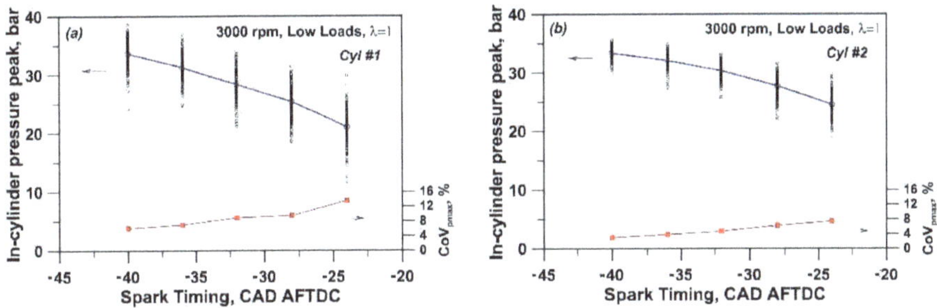

Figure 10. In-cylinder pressure peak and CoV pressure peak ($_{pmax}$) vs. Spark timing for Cyl#1 (**a**) and Cyl#2 (**b**) at 3000 rpm and low load.

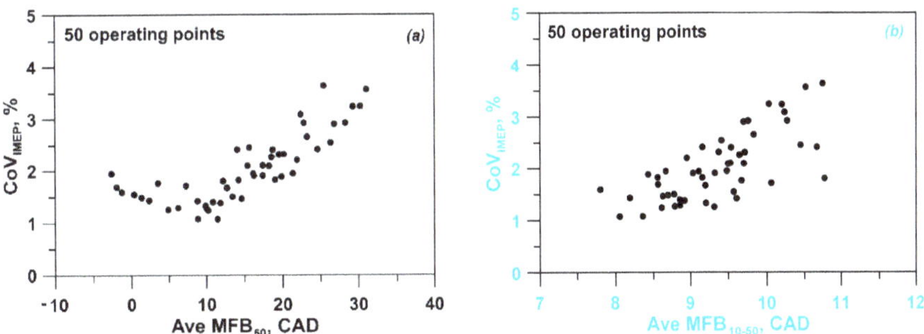

Figure 11. CoV_{IMEP} vs. averaged MFB_{50} (**a**) and MFB_{10-50} (**b**).

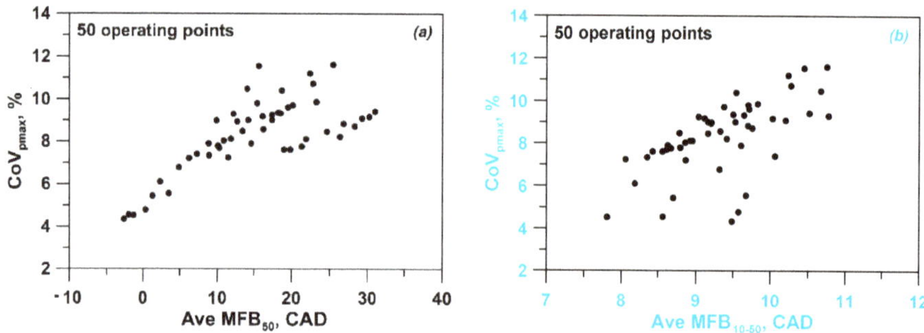

Figure 12. CoV_{pmax} vs. averaged MFB_{50} (**a**) and MFB_{10-50} (**b**).

Referring to the CoV_{IMEP}, higher values for delayed combustions are obtained (Figure 11a). A minimum level is also detected at a combustion phasing substantially corresponding to the one realizing the MBT condition. CoV_{IMEP} also shows a clear increasing trend with the combustion core duration MFB_{10-50} (Figure 11b). Concerning the CoV_{pmax}, it increases with a more delayed combustion process (Figure 12a), while a certain dispersed trend between p_{max} variation and combustion core duration is established in Figure 12b. However, a longer combustion phenomenon involves a greater CoV_{pmax}. The discussed experimental data on CCV in Figures 11 and 12 were utilized to define proper correlations with averaged combustion parameters through the 'curve fitting' module available in the Matlab code. In particular, for the examined engine, two polynomial correlations were identified by linking the CoV levels with the averaged combustion phasing and core duration:

$$CoV_{IMEP} = a_0 + a_1 MFB_{50} + a_2 MFB_{10-50} + a_3 MFB_{50}^2 + a_4 MFB_{50} MFB_{10-50} \quad (7)$$

$$CoV_{pmax} = b_0 + b_1 MFB_{50} + b_2 MFB_{10-50} + b_3 MFB_{50}^2 + b_4 MFB_{50} MFB_{10-50} \quad (8)$$

where a_i and b_i represent the constants for the best fit with the experimental outcomes. The proposed correlations (7) and (8) show a good coefficient of determination (Table 4).

Table 4. CoV correlation parameters.

CoV	i	0	1	2	3	4	R^2
CoV_{IMEP}	a_i	0.0902	−0.1586	0.1605	0.0021	0.0153	0.9236
CoV_{pmax}	b_i	8.869	−0.4873	−0.4674	−0.0147	0.1084	0.9442

The reliability of the experimental-based CoV correlations (7) and (8) was tested against the measured CoV values. The numerical/experimental comparisons for CoV of IMEP and p_{max} are reported in Figure 13: satisfactory agreements are observed for both correlations changing the speed/load points and at a null external EGR rate. A slightly better prediction is realized for the CoV_{IMEP}. In agreement with the experimental literature [35], Figure 13 confirms that IMEP cyclic dispersion is lower than the corresponding in-cylinder pressure peak cyclic dispersion.

Once validated, the developed CoV correlations were linked to the 1D computed combustion parameters (average cycle) in order to provide a valuable prediction of the cycle-to-cycle variation in the operating conditions not explored during the experimental activity. To this aim, the adoption of refined combustion and turbulence sub-models is mandatory. As extensively discussed in the following section, the main use of the developed correlations will be the prediction of the CCV level in the numerically optimized engine points, where the individual cylinders show the same nominal operation.

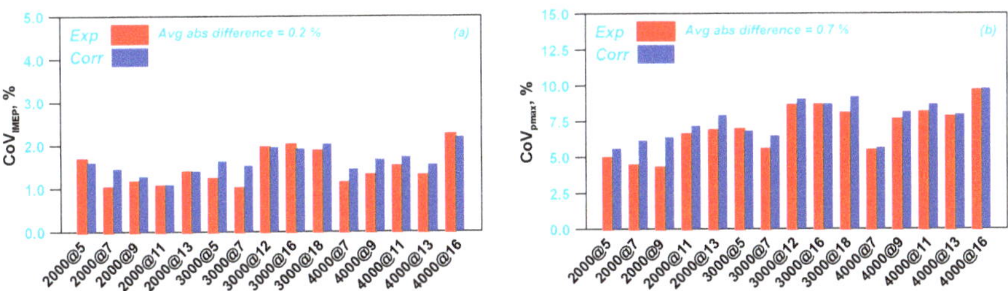

Figure 13. Experimental/Numerical comparison of Coefficient of Variation (CoV) of IMEP (**a**) and Coefficient of variation of in-cylinder pressure peak (**b**) at various engine operating points.

4. Numerical Optimization of Combustion: Results and Discussion

As already reported in previous sections, the operating points measured in the entire engine domain (Table 3) were considered to perform a combustion optimization, assuming the same behavior for each cylinder. This means that the numerical optimization process takes into account the absence of non-uniformities among cylinders, which practically requires only partial modifications to the existing fuel injection system of the original engine. The numerical optimization is performed by employing the validated 1D model and the validated CoV correlations referring to the examined engine. It consists of a virtual engine re-calibration, mainly acting on the fuel injection and on the spark timing. In particular, at each measured point in Table 3, a stoichiometric A/F ratio is imposed in each cylinder by a refined control of injected gasoline mass. This condition eliminates the experimental cylinder-to-cylinder variation. At low loads, a PID controller acting on the throttle valve opening is used to match the prescribed IMEP while the waste-gate valve is fully opened. At high loads, a PID controller acts on the waste-gate valve opening to match the target IMEP, while the throttle valve is considered fully opened. The spark timing is modified with the aim to identify the MBT condition and in the case of knock occurrence it is automatically delayed to maintain the numerical knock index below a prescribed threshold level. In addition, at low engine speeds a safety margin is applied to the maximum boost pressure to avoid the compressor surge [36]. At high loads, both the in-cylinder pressure peak and the temperature at turbine inlet are kept below maximum allowable levels to preserve the engine safety. The numerical results obtained by the model-related optimization (labelled as 'Opt') are compared to the model prediction based on the standard ECU calibration (labelled as 'Base'). The above comparisons are realized with the aim to highlight the potential benefits arising from the optimal operation of the engine cylinders. These numerical-derived advantages are here presented in terms of combustion phasing (MFB_{50}) and core duration (MFB_{10-50}), ISFC, and cyclic dispersion levels (CoV_{IMEP} and CoV_{pmax}). The following Figures 14–16 show the comparison of 'Base' and 'Opt' numerical solutions for three different engine speeds (2000, 3000 and 4000 rpm) and at various IMEP values. For the knock-free part load points at 2000 rpm (below 9 bar IMEP), the model does not substantially improve the combustion process with respect to the Base combustion parameters (Figure 14a,b). The latter are taken as the average of combustion variables between the two cylinders. The ISFC improvements at low loads (Figure 14c) are exclusively to be attributed to the suppression of the A/F ratio unbalance between cylinders and a maximum percent gain of 5.6% at 2000@5 is reached. Conversely, at higher loads under knock-limited operation (2000@11 and 2000@13), a combustion phasing advance is realized (Figure 14a), allowing a shortening of combustion core duration (Figure 14b). ISFC benefits at high loads (2.2% for both 2000@11 and 2000@13 in Figure 14c) have to be ascribed to an improved combustion process under in-cylinder stoichiometric mixture conditions. Obviously, the combustion optimization at high load also reflects on the reduction of CoV levels (Figure 14d), while at knock-free loads the CCV remains unchanged. The numerical optimization outcomes at speeds of 3000 and 4000 rpm show similarities. Indeed, for both speeds and under knock-limited operations (above 12 bar IMEP at 3000 rpm and 11 bar IMEP at 4000 rpm), the identified optimal combustion phasing is advanced (Figure 15a/Figure 16a) with respect to the Base level and, consequently, the combustion core duration is also reduced (Figure 15b/Figure 16b). Lower CoV levels are predicted by the developed CCV correlations for the optimal solutions, mainly thanks to the improved combustion process (Figure 15d/Figure 16d). Concerning the ISFC advantages, the previous considerations for the case at low speed can be also applied for the medium and high speeds. It is worth highlighting that greater ISFC percent gains are realized in all the high load points where the ECU standard calibration required a rich combustion (Table 3). As confirmed in Figure 15c, higher ISFC benefits are achieved at 3000@12, 3000@16 and 3000@18 with gains of 9.7%, 13.3% and 12.2%, respectively. In a similar way, Figure 16c shows a very high ISFC advantage (equal to 14.7%) only at 4000@16. Finally, at knock-free loads and medium/high speeds (3000@5, 3000@7, 4000@7 and 4000@9) limited improvements in the combustion process and CoV indexes are achieved through the optimization. In these cases, the fuel consumption is still improved (ISFC gain of 4.4% at 3000@5 in Figure 15c and 3.5% at 4000@7 in Figure 16c), due to the elimination of an uneven A/F ratio between cylinders. Summarizing, the suppression of cylinder-to-cylinder variation through the optimization of single cylinder operation demonstrated the possibility of achieving improvements in terms of combustion evolution and stability also contributing to reduced fuel consumption while preserving engine safety both in terms of mechanical and thermal stresses.

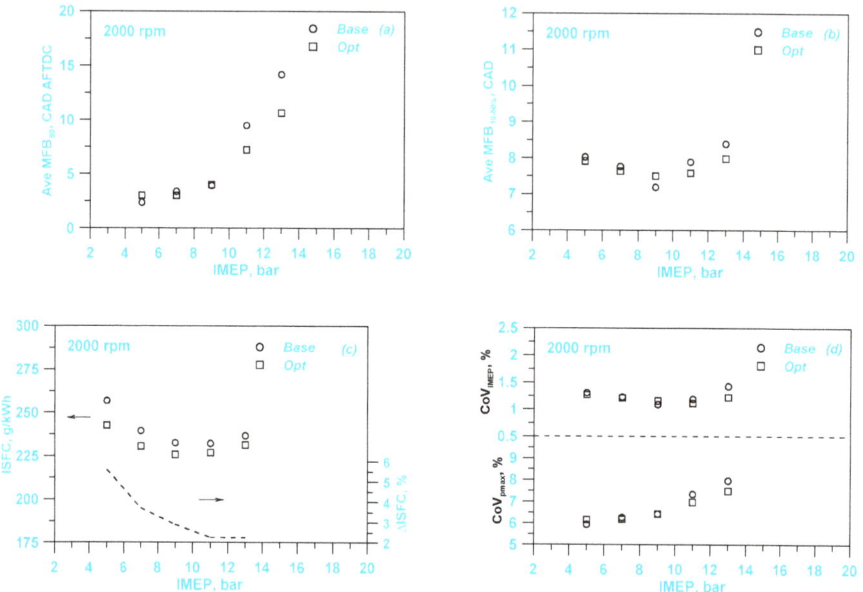

Figure 14. Comparison of Base and Optimized solutions for different engine operating points in a IMEP sweep at constant speed of 2000 rpm: Ave. MFB_{50} (**a**), Ave. MFB_{10-50} (**b**), ISFC (**c**), CoV_{IMEP} and CoV_{pmax} (**d**).

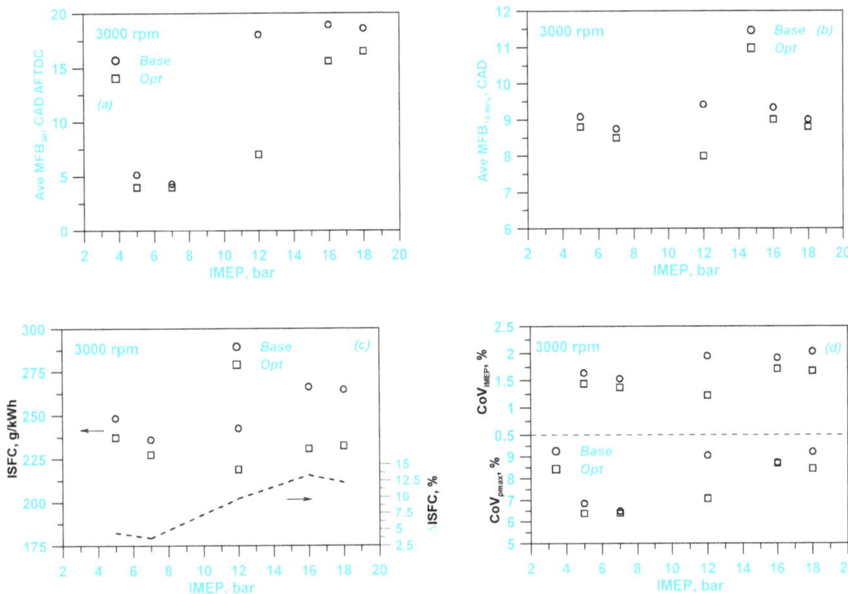

Figure 15. Comparison of Base and Optimized solutions for different engine operating points in a IMEP sweep at constant speed of 3000 rpm: Ave. Ave. MFB_{50} (**a**), Ave. $MFB_{10\text{–}50}$ (**b**), ISFC (**c**), CoV_{IMEP} and CoV_{pmax} (**d**).

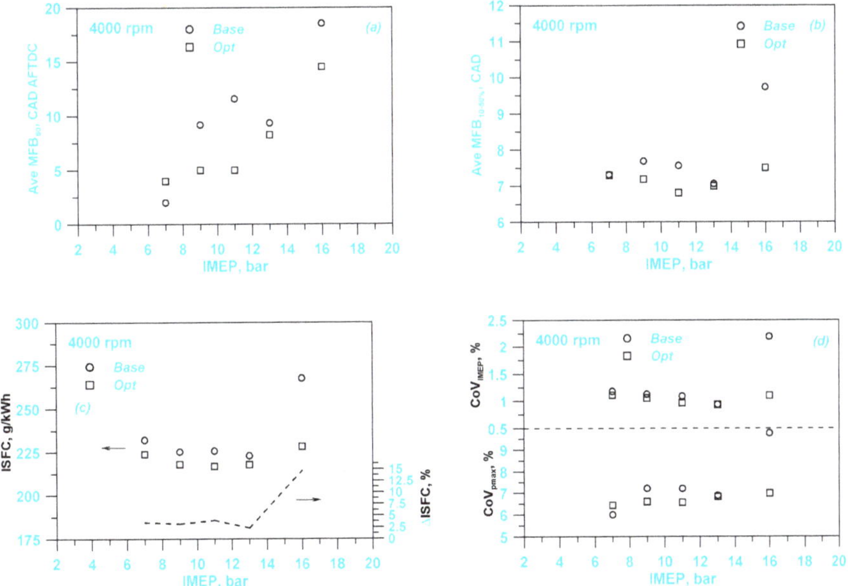

Figure 16. Comparison of Base and Optimized solutions for different engine operating points in a IMEP sweep at constant speed of 4000 rpm: Ave. MFB_{50} (**a**), Ave. $MFB_{10\text{–}50}$ (**b**), ISFC (**c**), CoV_{IMEP} and CoV_{pmax} (**d**).

5. Conclusions

In this work, a small, turbocharged Spark Ignition engine is numerically and experimentally studied with the aim to optimize the individual cylinder combustion and, consequently, to reduce the engine fuel consumption and the cycle-to-cycle variation. In a first phase two different experimental steady analyses were carried out. In particular, a spark advance sweep was performed both at high and low loads, including various engine speeds and air/fuel mixture qualities. The acquired data were post-processed to find proper engine CoV correlations of IMEP and in-cylinder pressure peak with the average combustion phasing and core duration. A second experimental investigation was realized, mainly consisting of a load sweep at three different speeds (namely 2000, 3000 and 4000 rpm), covering both knock-free and knock-limited operations in the engine domain. The operating points considered in this phase of experiments correspond to the conditions imposed by the ECU in the commercial vehicle. The engine was schematized in a 1D model which was validated in terms of global performance variables, combustion evolution and average pressure cycles obtained from the above experimental IMEP sweep. Then, the developed CoV correlations were validated against the experimentally derived CoV values of measured IMEP-sweep points. Once validated, 1D model and the experimentally-related CoV correlations were coupled with the aim to realize a numerical optimization of the engine performance under knock-free and knock-limited conditions. The optimization process reveals that the individual cylinder operation can be considerably improved from ECU-defined settings, especially at high loads. The indicated specific fuel consumption is reduced if compared to the experimental calibration and an ISFC percent benefit ranging from 2.2% up to 14.7% is attained at varying the speed/load points. At low loads, the optimized ISFC level is reached mainly thanks to the possibility of stoichiometric A/F ratio operation in each cylinder. At high loads, ISFC is numerically optimized through the definition of a stoichiometric A/F ratio in each cylinder and an improvement in the combustion process. The optimization at high loads allows the definition of a new setting for the knock-limited spark advance (KLSA) which implies an advanced combustion phasing (MFB_{50}) and a reduced combustion duration if compared to the experimental ECU-defined counterparts. Furthermore, the validated CoV correlations, starting from the numerically optimized combustion parameters, furnish an empirical-based prediction of the cycle-to-cycle variation improvement. Consistently with the combustion optimization, significant reduction in the cycle-to-cycle variation is achieved at high loads for each investigated speed. Summarizing, the adopted numerical procedure allows the identification of the optimal combustion operation, contributing to support the control of each cylinder operation. In this way, the proposed integrated numerical models represent a useful and valuable tool to limit the fuel consumption, the cycle-to-cycle and the cylinder-to-cylinder variations during the engine development phase.

Author Contributions: Conceptualization, L.M.; methodology, C.T., L.M. and L.T.; software, L.T.; validation, L.T.; formal analysis, C.T., L.M. and L.T.; investigation, C.T., L.M. and L.T.; data curation, C.T., L.M. and L.T.; writing—original draft preparation, C.T. and L.T.; writing—review and editing, C.T., L.M. and L.T. All authors have read and agreed to the published version of the manuscript.

Funding: This research received no external funding.

Acknowledgments: The authors are grateful to Alfredo Mazzei (Istituto Motori, CNR) for the technical support that has been fundamental for the preparation of the engine test bench.

Conflicts of Interest: The authors declare no conflict of interest.

Abbreviations

0D	Zero Dimensional
1D	One Dimensional
η	Efficiency
λ	Relative air-fuel ratio
ρ	Density

Σ	Turbulence-induced flame wrinkling
τ_{AI}	Auto-ignition Time
A	Area
A/F	Air/Fuel
AFTDC	After Top Dead Center
AI	Auto Ignition
A_L	Laminar Flame Area
A_s	Heat transfer surface area
BSFC	Brake Specific Fuel Consumption
CAD	Crank Angle Degree
CCV	Cycle-to-Cycle Variation
C_f	Friction coefficient
Cov	Coefficients of Variation
D	Equivalent diameter
dp	Pressure differential across dx
dx	Discretization length
e	Specific internal energy
e-EGR	external Exhaust Gas Recirculation
ECU	Engine Control Unit
EVC	Exhaust Valve Closure
EVO	Exhaust Valve Opening
FEM	Finite Element Method
$FMEP$	Friction Mean Effective Pressure
h	Heat transfer coefficient
H	Specific enthalpy
IMEP	Indicated Mean Effective Pressure
ISFC	Indicated Specific Fuel Consumption
IVC	Intake Valve Closure
IVO	Intake Valve Opening
k	Turbulent kinetic energy
K	Mean flow kinetic energy
K_p	Pressure loss coefficient
LFS	Laminar Flame Speed
m	mass
m_b	Burned mass
$m_{b,entr}$	Burned portion of entrained mass
MBT	Maximum Brake Torque
m_{entr}	Current entrained mass
MFB	Mass Fraction Burned
$\dot{m}$	Mass flow Rate
$\dot{m}_f$	Total Fuel Flow Rate
n	Engine rotational speed
p	Pressure
P	Power
PFI	Port Fuel Injection
PID	Proportional Integral Derivative
SI	Spark Ignition
S_L	Laminar Flame Speed
t	Time
T	Tumble angular momentum
T_{fluid}	Fluid Temperature
T_{wall}	Wall Temperature
u	Velocity at boundary
V	Volume
VVA	Variable Valve Actuation
WLTCSI	Worldwide harmonized Light vehicles Test Cycle

References

1. Gavhane, R.S.; Kate, A.M.; Pawar, A.; Safaei, M.R.; Soudagar, M.E.M.; Mujtaba, M.; Ali, H.M.; Banapurmath, N.R.; Goodarzi, M.; Badruddin, I.A.; et al. Effect of Zinc Oxide Nano-Additives and Soybean Biodiesel at Varying Loads and Compression Ratios on VCR Diesel Engine Characteristics. *Symmetry* **2020**, *12*, 1042. [CrossRef]
2. Khan, H.; Soudagar, M.E.M.; Kumar, R.H.; Safaei, M.R.; Farooq, M.; Khidmatgar, A.; Banapurmath, N.R.; Farade, R.A.; Mujtaba, M.; Afzal, A.; et al. Effect of Nano-Graphene Oxide and n-Butanol Fuel Additives Blended with Diesel—Nigella sativa Biodiesel Fuel Emulsion on Diesel Engine Characteristics. *Symmetry* **2020**, *12*, 961. [CrossRef]
3. Wirth, M.; Schulte, H. Downsizing and Stratified Operation—An Attractive Combination Based on a Spray-guided Combustion System. *Fortschr. Ber. VDI REIHE 12 Verk. Fahrz.* **2006**, *622*, 68.
4. Fontana, G.; Galloni, E. Variable valve timing for fuel economy improvement in a small spark-ignition engine. *Appl. Energy* **2009**, *86*, 96–105. [CrossRef]
5. Kumano, K.; Yamaoka, S. Analysis of Knocking Suppression Effect of Cooled EGR in Turbo-Charged Gasoline Engine. *SAE Tech. Pap.* **2014**. [CrossRef]
6. Tornatore, C.; Bozza, F.; De Bellis, V.; Teodosio, L.; Valentino, G.; Marchitto, L. Experimental and numerical study on the influence of cooled EGR on knock tendency, performance and emissions of a downsized spark-ignition engine. *Energy* **2019**, *172*, 968–976. [CrossRef]
7. Boretti, A. Water injection in directly injected turbocharged spark ignition engines. *Appl. Therm. Eng.* **2013**, *52*, 62–68. [CrossRef]
8. Masouleh, M.G.; Keskinen, K.; Kaario, O.; Kahila, H.; Wright, Y.M.; Vuorinen, V. Flow and thermal field effects on cycle-to-cycle variation of combustion: Scale-resolving simulation in a spark ignited simplified engine configuration. *Appl. Energy* **2018**, *230*, 486–505. [CrossRef]
9. Finney, C.E.; Kaul, B.C.; Daw, C.S.; Wagner, R.M.; Edwards, K.D.; Green, J.J.B. Invited Review: A review of deterministic effects in cyclic variability of internal combustion engines. *Int. J. Engine Res.* **2015**, *16*, 366–378. [CrossRef]
10. Ozdor, N.; Dulger, M.; Sher, E. *Cyclic Variability in Spark Ignition Engines a Literature Survey*; SAE Technical Paper 940987; SAE International: Warrendale, PA, USA, 1994.
11. Mahdisoozani, H.; Mohsenizadeh, M.; Bahiraei, M.; Kasaeian, A.; Daneshvar, A.; Goodarzi, M.; Safaei, M.R. Performance Enhancement of Internal Combustion Engines through Vibration Control: State of the Art and Challenges. *Appl. Sci.* **2019**, *9*, 406. [CrossRef]
12. Saxén, J.-E.; Hyvämäki, T.; Björkqvist, J.; Ostman, F.; Toivonen, H.T. Power Balancing of Internal Combustion Engines—A Time and Frequency Domain Analysis. *IFAC Proc. Vol.* **2014**, *47*, 10802–10807. [CrossRef]
13. Zhou, F.; Fu, J.; Shu, J.; Liu, J.; Wang, S.; Feng, R. Numerical simulation coupling with experimental study on the non-uniform of each cylinder gas exchange and working processes of a multi-cylinder gasoline engine under transient conditions. *Energy Convers. Manag.* **2016**, *123*, 104–115. [CrossRef]
14. Baratta, M.; D'Ambrosio, S.; Misul, D.A.; Spessa, E. Effects of H2 Addition to Compressed Natural Gas Blends on Cycle-to-Cycle and Cylinder-to-Cylinder Combustion Variation in a Spark-Ignition Engine. *J. Eng. Gas Turbines Power* **2014**, *136*, 051502. [CrossRef]
15. Maiboom, A.; Tauzia, X.; Hétet, J.-F. Influence of EGR unequal distribution from cylinder to cylinder on NOx–PM trade-off of a HSDI automotive Diesel engine. *Appl. Therm. Eng.* **2009**, *29*, 2043–2050. [CrossRef]
16. Persson, H.; Pfeiffer, R.; Hultqvist, A.; Johansson, B.; Ström, H. *Cylinder-to-Cylinder and Cycle-to-Cycle Variations at HCCI Operation with Trapped Residuals*; SAE Technical Paper 2005-01-0130; SAE International: Warrendale, PA, USA, 2005. [CrossRef]
17. Luján, J.M.; Galindo, J.; Serrano, J.; Pla, B. A methodology to identify the intake charge cylinder-to-cylinder distribution in turbocharged direct injection Diesel engines. *Meas. Sci. Technol.* **2008**, *19*, 065401–065411. [CrossRef]
18. Hyvonen, J.; Haraldsson, G.; Johansson, B. *Balancing Cylinder-to-Cylinder Variations in a Multi-Cylinder VCR-HCCI Engine*; SAE Technical Paper 2004-01-1897; SAE International: Warrendale, PA, USA, 2004. [CrossRef]

19. Yun, H.; Kang, J.-M.; Chang, M.-F.; Najt, P. *Improvement on Cylinder-to-Cylinder Variation Using a Cylinder Balancing Control Strategy in Gasoline HCCI Engines*; SAE Technical Paper 2010-01-0848; SAE International: Warrendale, PA, USA, 2010. [CrossRef]
20. Xu, Z.; Zhang, Y.; Di, H.; Shen, T. Combustion variation control strategy with thermal efficiency optimization for lean combustion in spark-ignition engines. *Appl. Energy* **2019**, *251*, 113329. [CrossRef]
21. Naruke, M.; Morie, K.; Sakaida, S.; Tanaka, K.; Konno, M. Effects of hydrogen addition on engine performance in a spark ignition engine with a high compression ratio under lean burn conditions. *Int. J. Hydrogen Energy* **2019**, *44*, 15565–15574. [CrossRef]
22. Yu, X.; Wu, H.; Du, Y.; Tang, Y.; Liu, L.; Niu, R. Research on cycle-by-cycle variations of an SI engine with hydrogen direct injection under lean burn conditions. *Appl. Therm. Eng.* **2016**, *109*, 569–581. [CrossRef]
23. Masouleh, M.G.; Keskinen, K.; Kaario, O.; Kahila, H.; Karimkashi, S.; Vuorinen, V. Modeling cycle-to-cycle variations in spark ignited combustion engines by scale-resolving simulations for different engine speeds. *Appl. Energy* **2019**, *250*, 801–820. [CrossRef]
24. Selim, M.Y. Sensitivity of dual fuel engine combustion and knocking limits to gaseous fuel composition. *Energy Convers. Manag.* **2004**, *45*, 411–425. [CrossRef]
25. Bozza, F.; Gimelli, A.; Merola, S.S.; Vaglieco, B. *Validation of a Fractal Combustion Model through Flame Imaging*; SAE Technical Paper 2005-01-1120; SAE International: Warrendale, PA, USA, 2005.
26. Gouldin, F. An application of fractals to modeling premixed turbulent flames. *Combust. Flame* **1987**, *68*, 249–266. [CrossRef]
27. Bozza, F.; Teodosio, L.; De Bellis, V.; Fontanesi, S.; Iorio, A. A Refined 0D Turbulence Model to Predict Tumble and Turbulence in SI Engines. *SAE Int. J. Engines* **2018**, *12*, 15–30. [CrossRef]
28. Teodosio, L.; Pirrello, D.; Berni, F.; De Bellis, V.; Lanzafame, R.; D'Adamo, A. Impact of intake valve strategies on fuel consumption and knock tendency of a spark ignition engine. *Appl. Energy* **2018**, *216*, 91–104. [CrossRef]
29. Teodosio, L.; Bozza, F.; Tufano, D.; Giannattasio, P.; Distaso, E.; Amirante, R. Impact of the laminar flame speed correlation on the results of a quasi-dimensional combustion model for Spark-Ignition engine. *Energy Procedia* **2018**, *148*, 631–638. [CrossRef]
30. Teodosio, L. Modeling of Turbulence, Combustion and Knock for Performance Prediction, Calibration and Design of a Turbocharged Spark Ignition Engine. Ph.D. Thesis, Industrial Engineering Department of University of Naples "Federico II", Naples, Italy, May 2016. Available online: http://www.fedoa.unina.it/id/eprint/10725 (accessed on 6 May 2016).
31. Bozza, F.; De Bellis, V.; Teodosio, L. *A Tabulated-Chemistry Approach Applied to a Quasi-Dimensional Combustion Model for a Fast and Accurate Knock Prediction in Spark-Ignition Engines*; SAE Technical Paper 2019-01-0471; SAE International: Warrendale, PA, USA, 2019.
32. Andrae, J.C.G.; Kovács, T. Evaluation of Adding an Olefin to Mixtures of Primary Reference Fuels and Toluene To Model the Oxidation of a Fully Blended Gasoline. *Energy Fuels* **2016**, *30*, 7721–7730. [CrossRef]
33. Hohenberg, G.F. *Advanced Approaches for Heat Transfer Calculations*; SAE Transactions: Warrendale, PA, USA, 1979; pp. 2788–2806.
34. Teodosio, L.; Bozza, F.; Berni, F. Effects of nanofluid contaminated coolant on the performance of a spark ignition engine. *AIP Conf. Proc.* **2019**, 020147. [CrossRef]
35. De Bellis, V.; Teodosio, L.; Siano, D.; Minarelli, F.; Cacciatore, D. Knock and Cycle by Cycle Analysis of a High Performance V12 Spark Ignition Engine. Part 1: Experimental Data and Correlations Assessment. *SAE Int. J. Engines* **2015**, *8*, 1993–2001. [CrossRef]
36. Bozza, F.; De Bellis, V.; Teodosio, L.; Gimelli, A. Numerical analysis of the transient operation of a turbocharged diesel engine including the compressor surge. *Proc. Inst. Mech. Eng. Part D J. Automob. Eng.* **2013**, *227*, 1503–1517. [CrossRef]

Publisher's Note: MDPI stays neutral with regard to jurisdictional claims in published maps and institutional affiliations.

Article

Arc-Phase Spark Plug Energy Deposition Characteristics Measured Using a Spark Plug Calorimeter Based on Differential Pressure Measurement

Kyeongmin Kim, Matthew J. Hall *, Preston S. Wilson and Ronald D. Matthews

Department of Mechanical Engineering, The University of Texas at Austin, Austin, TX 78712, USA; kyekim@utexas.edu (K.K.); pswilson@mail.utexas.edu (P.S.W.); rdmatt@mail.utexas.edu (R.D.M.)
* Correspondence: mjhall@mail.utexas.edu; Tel.: +1-512-758-1361

Received: 3 June 2020; Accepted: 7 July 2020; Published: 10 July 2020

Abstract: A spark plug calorimeter is introduced for quantifying the thermal energy delivered to unreactive gas surrounding the spark gap during spark ignition. Unlike other calorimeters, which measure the small pressure rise of the gas above the relatively high gauge pressure or relative to an internal reference, the present calorimeter measured the differential rise in pressure relative to the initial pressure in the calorimeter chamber. By using a large portion of the dynamic range of the chip-based pressure sensor, a high signal to noise ratio is possible; this can be advantageous, particularly for high initial pressures. Using this calorimeter, a parametric study was carried out, measuring the thermal energy deposition in the gas and the electrical-to-thermal energy conversion efficiency over a larger range of initial pressures than has been carried out previously (1–24 bar absolute at 298 K). The spark plug and inductive ignition circuit used gave arc-type rather than glow-type discharges. A standard resistor-type automotive spark plug was tested. The effects of spark gap distance (0.3–1.5 mm) and ignition dwell time (2–6 ms) were studied for an inductive-type ignition system. It was found that energy deposition to the gas (nitrogen) and the electrical-to-thermal energy conversion efficiency increased strongly with increasing gas pressure and spark gap distance. For the same ignition hardware and operating conditions, the thermal energy delivered to the gap varied from less than 1 mJ at 1 atm pressure and a gap distance of 0.3 mm to over 25 mJ at a pressure of 24 bar and a gap distance of 1.5 mm. For gas densities that might be representative of those in an engine at the time of ignition, the electrical-to-thermal energy conversion efficiencies ranged from approximately 3% at low pressures (4 bar) and small gap (0.3 mm) to as much as 40% at the highest pressure of 24 bar and with a gap of 1.5 mm.

Keywords: spark ignition; calorimeter; thermal energy; spark plug; natural gas engine

---

1. Introduction

Spark ignition engines rely on the thermal energy deposited by the spark plasma into the in-cylinder gases to initiate combustion each engine cycle. The thermal energy density must be sufficient to initiate and sustain the nascent flame kernel until the chemical heat release from the flame kernel is enough to create a self-sustaining pre-mixed turbulent flame that propagates throughout the combustion chamber on a millisecond time scale. The reliability of the ignition process depends on many factors including fuel type, fuel–air mixture ratio, mixture homogeneity, dilution (e.g., exhaust gas recirculation (EGR) level), residual fraction, in-cylinder bulk flows and turbulence, spark plug geometry, spark timing and duration, and the characteristics of the electrical energy delivery [1]. These factors can impact the amount of thermal energy deposition to the gas from the plasma and its distribution in the vicinity of the spark gap. High dilution rates and lean mixtures pose especially challenging conditions for achieving reliable combustion initiation that minimizes cycle-to-cycle variation.

Ever more detailed ignition submodels that couple to engine computational fluid dynamics (CFD) simulation codes such as Converge CFD® are being developed [2–4]. They are increasingly based on spatially-resolved spark gap physics and can include coupling to dynamic ignition circuit models. Validation of the electrical-to-thermal energy conversion and gas-to-electrode heat transfer rates is needed for these models.

For the above reasons, it is important to understand the characteristics of the thermal energy deposition from the engine spark plasma to the gases. Many fundamental studies of spark breakdown behavior have been conducted. Some of these studies are summarized in Meeks [5]. A wealth of information and data characteristic of the behavior of spark ignition systems of the type used in engines can be found in Maly [6]. By measuring the ignition voltage and current characteristics at the spark plug, it is possible to determine the electrical energy delivered to the spark gap [7], but that does not quantify the thermal energy deposited in the gas in the gap that is responsible for flame initiation. For this purpose, spark plug calorimeters have been developed that will measure the thermal energy deposition. These calorimeters measure the pressure rise of the gas in a small chamber in which the spark plug is located to determine the thermal energy input to the gas.

This work reports on the development of a spark plug calorimeter that uses very sensitive interchangeable pressure sensors to measure the differential pressure rise of the chamber gases relative to the initial chamber pressure and is designed to function at very high pressures (>20 bar). The calorimeter was then used to investigate the thermal energy deposition and efficiency of electrical-to-thermal energy conversion of arc-type spark plug discharges for an inductive ignition system for different pressures, gap distances and dwell times.

High pressure measurements are of increasing interest due to the current push by many engine manufacturers to develop ever higher brake mean effect pressure (BMEP) engines that rely on increasingly high boost pressures. Most previous studies that have used spark calorimeters have relied upon pressure sensors that measured the spark plug chamber gauge pressure relative to the ambient atmospheric pressure or are designed for use at high pressures and measure relative to an internal reference. By measuring the differential chamber pressure relative to the initial pressure, more sensitive measurements with a higher dynamic range are possible.

Several studies that used spark plug calorimeters are found in the literature. Roth et al. [8] first introduced the measurement of spark energy delivered to the gas with both constant-volume and constant-pressure calorimeters. They performed a fundamental study of the effects of the electrode diameter, gap distance, and thermal diffusivity of the gas on the electrical-to-thermal energy conversion efficiency with monatomic gases. They found that more of the electrical energy from the spark was lost to the electrodes with larger electrode diameter, shorter gap distance, and gases with larger thermal diffusivity. Merritt [9] was first to create a spark calorimeter consisting of two chambers and a differential pressure transducer. It was constructed of plastic and was used to measure spark energy thermal deposition at atmospheric pressure. The extra reference volume was said to eliminate the short-term fluctuations in atmospheric pressure. This differential pressure sensing concept for a spark calorimeter was adopted by Franke and Reinmann [10]. They measured the energy delivered to the gas comparing five different ignition systems that included both capacitive and inductive ignition types at pressures up to 16 bar; all of the systems delivered breakdown followed by glow discharges, without a discernable arc phase. The mean electrical-to-thermal energy conversion efficiency increased from 5% to 50% as the pressure varied from 1 to 16 bar, and for a given level of delivered electrical energy, the efficiency increased as the duration of the discharge increased.

While the concept of a spark plug calorimeter based on differential pressure measurement is the same here as that presented by Franke and Reinmann, the present paper goes beyond their work in several significant ways, both in terms of the experimental technique and the conditions analyzed. With respect to the calorimeter design, one difference is that in this study we used inexpensive and easily changeable solid-state semiconductor pressure sensors with different sensitivities such that we could exchange them to adapt to different expected pressure rises, allowing us to maximize the dynamic range of the measurements. Another novel feature of this calorimeter is the system that allowed the pressure sensor to be calibrated at high pressure for very small pressure differentials. The relatively small calorimeter volume resulted in greater pressure rises for a given thermal energy deposition, further increasing the potential accuracy of the measurements.

With respect to the differences in experimental conditions, the present study extended the pressure range of the measurements by 50% relative to that of the study by Franke and Reinmann. This is especially significant for application to advanced natural gas engines under development by various engine manufacturers. These engines are designed for ever higher boost conditions and high BMEPs that result in ever higher gas densities at the time of ignition. In addition, this study is primarily focused on arc-type discharges. Franke and Reinmann stated that "In this experiment, no transition to arc mode has been observed", implying that they believed that the plasma discharges that they observed were in the glow regime. Further, in this study, we present empirically-based analytical expressions for how differential or incremental changes in thermal energy deposition scale with changes in either gap distance or gas pressure.

Other spark calorimeter studies found in the literature used calorimeters that measured the chamber pressure relative to the outside ambient pressure, a simpler method, but one that compromises sensitivity and dynamic range, particularly at high initial pressures. Teets and Sell [11] used such a calorimeter to study the thermal energy deposition characteristics of three different ignition systems that included an inductive system, a plasma jet ignitor and an ultra-short pulse ignitor. Measurements were to pressures up to 7 bar. They found conversion efficiencies varied from 5% to 65%, increasing with pressure and gap distance. They found that electrical-to-thermal energy conversion efficiency decreased as the delivered electrical energy increased, and also found the efficiency of the ultra-short pulse system higher than the inductive system.

Verhoeven [12] took a different approach to measure thermal energy deposition to the gas from a spark plasma, using an optical holographic technique. Uncertainties associated with the technique were greater than for calorimeters, but it had the advantage of allowing measurements in flowing gases. However, the flow measurements were only made at a pressure of 1 atm. They found that with a cross-flow velocity of 5 m/s, energy deposition efficiency was approximately three-fold that for a quiescent gas.

In a limited study, Abidin et al. [13] investigated the effect of dwell time, gap distance, and pressure on the breakdown voltage and the electrical-to-thermal energy conversion efficiency using a standard spark calorimeter. The electrical-to-thermal efficiency increased with larger gap distance, shorter dwell time, and higher pressure. However, the highest pressure investigated was 9 bar. Alger et al. [14] used a spark calorimeter to measure the energy delivered to the gas while investigating the effect of spark plug design on the initial flame kernel development. They found typical electrical-to-thermal energy conversion efficiencies of approximately 20% and found that spark plugs with high internal resistances had higher thermal energy depositions, which resulted in faster flame kernel development.

In this study, a spark plug calorimeter was used to measure the thermal energy delivered to the nitrogen gas surrounding the spark gap during spark ignition. Unlike most other spark plug calorimeters, which measure the small pressure rise of the gas above the relatively high gauge pressure or relative to an internal reference, the present calorimeter measured the differential rise in pressure relative to the initial pressure in the calorimeter chamber. Using this calorimeter, a study was carried out, measuring the thermal energy deposition in the gas and the electrical-to-thermal energy conversion efficiency over a larger range of initial pressures than has been carried out previously. The measurements were made at pressures up to 24 bar. A pressure of 24 bar at the ambient temperature at which the measurements were made (approximately 22 °C) results in gas densities equivalent to approximately 50 bar for a gas temperature at the time of ignition of 600 K or 70 bar at 900 K. The spark plug and inductive ignition circuit used gave arc-type rather than glow-type discharges.

For example, for a reasonable intercooler exit temperature of 350 K, with this also taken as the intake temperature, and assuming a compression ratio of 10 while assuming polytropic compression with a typical polytropic exponent of 1.33, yields a TDC temperature of approximately 750 K. An ignition timing at or near TDC is often utilized in modern SI engines to enhance catalyst heating after startup.

2. Calorimeter Design

A photo of the calorimeter is shown in Figure 1a, along with a cross-sectional drawing shown in Figure 1b. It was machined from stainless steel in two pieces and accommodates a 14 mm spark plug. The calorimeter has two chambers, a small cylindrical chamber into which the spark plug is inserted and a second chamber in which the chip-based pressure sensor is located. The pressure sensor is mounted through a small hole that connects the two chambers. Two valves are used. One valve is used to fill the two chambers with gas (with both valves open). That valve is then closed to stop the flow of gas into the chambers. The second valve is then closed, which isolates the two chambers except for the small hole between them into which the pressure sensor entrance passage is sealed. The spark plug chamber is cylindrical in shape, with a diameter of 12.8 mm, a height of 9.5 mm, and a volume of 1.8 cm^3. The pressure sensor is sealed within the second cavity of the calorimeter. A high pressure electrical feedthrough is passed the necessary wires for signal and power source to the outside.

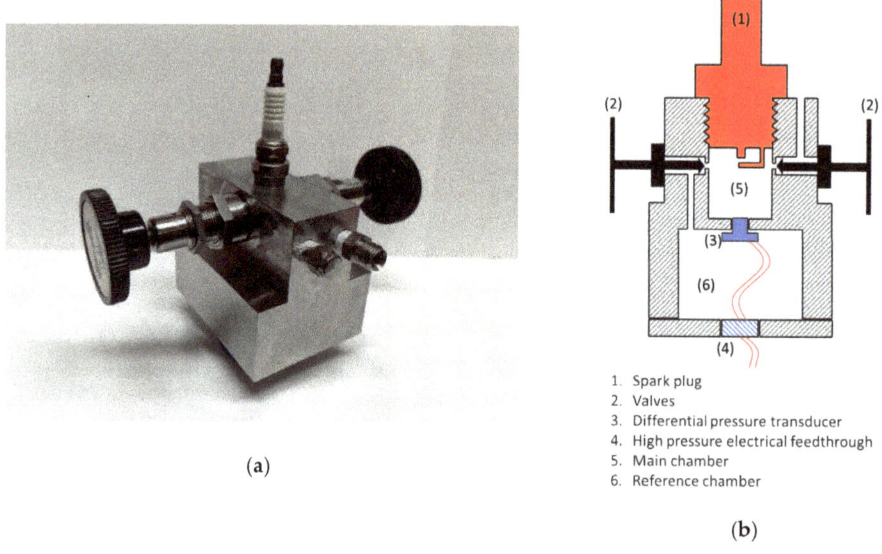

Figure 1. (**a**) Photo of spark calorimeter; (**b**) section drawing of the calorimeter.

The pressure sensors used were board mounted integrated silicon pressure sensors. Two types were used for the measurements presented here—one with a 3.9 kPa pressure range and another with a 10 kPa pressure range—depending on the expected pressure levels. The model numbers of the two sensors available from Freescale (USA) were NXP MP3V5004GC6U and MP3V5010GC6U for the 3.9 and 10 kPa sensors respectively. The 1/e response time of the sensors was approximately 0.2 ms. This was fast enough to quantify the trends in the cumulative energy deposition of an ignition event, but was not fast enough to resolve the thermal energy deposition associated with the breakdown process, which occurs on a submicrosecond time scale.

Figure 2 shows a schematic of the calorimeter experimental setup. For the measurements, the calorimeter was pressurized with nitrogen gas to the desired level. The charging pressure was measured using precision Bourdon tube pressure gauges. Time-resolved measurements of spark plug voltage and current were taken to determine the electrical energy delivered to the spark plug. A Tektronix Model P6015A high voltage probe measured the breakdown and follow-on voltages at the top of the spark plug. The current-dependent resistance of the plug was determined and the IR voltage subtracted from the voltage measured at the top of the plug to obtain the gap voltage. A Pearson Model 110 current sensor was used to measure the discharge current. The voltage, current, and pressure sensor signals were recorded using a 100 MHz 4-channel Tektronix oscilloscope. Breakdown voltages were recorded separately since a faster time-base setting was needed to resolve these very short duration events.

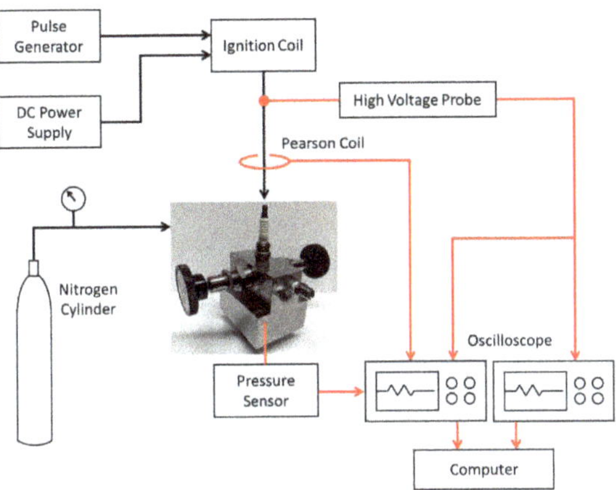

Figure 2. Schematic of calorimeter setup.

The chip-based sensors used to measure the change in chamber pressure that resulted from the spark discharge were intended, by the manufacturer, to measure small differential pressures relative to the ambient barometric pressure. Since the intention was to make measurements at high pressure, the sensors needed to be tested for their response at high pressure. To do this, a calibration procedure was developed that tested the pressure sensors in situ. The calibration device is depicted in Figure 3. A glass pipette was inserted through the spark plug hole and coupled to the passage tube of the pressure sensor with a short piece of plastic tubing. The pipette was then filled with water to a water column height that gave the desired pressure differential, the pipette thus acting as a manometer. A closed-end stainless steel tube was then fit over the pipette and secured with tube fittings. The two chambers of the calorimeter were when charged with nitrogen to the same gas pressure; the difference in pressure between the two chambers at the pressure sensor was the pressure imposed by the vertical column of water. This method allowed the pressure sensors to be calibrated for different high initial pressures and provided pressure measurements relative to the initial pressure.

The calibration data for the 10 kPa range sensor is shown in Figure 4, which shows the measured voltage output of the pressure sensor versus the differential pressure imposed by the water column height for initial total gas pressures from 1 atm to 24 bar. The response of the pressure sensor is shown over the measured range; there was no systematic differential response to the initial pressure. The scatter/reproducibility of the calibration data provides an estimate of the uncertainty in the accuracy of the pressure measurements of less than ± 0.3 kPa.

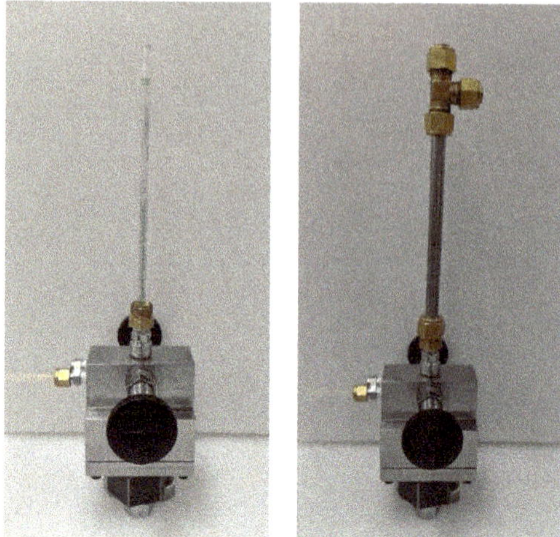

Figure 3. Photo of calorimeter with calibration tube attached.

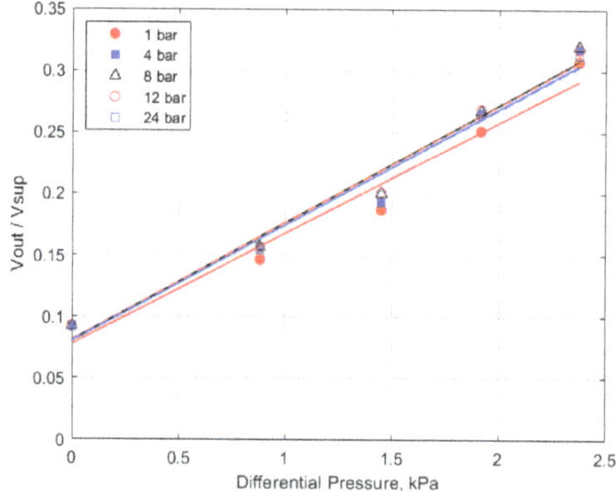

Figure 4. Calibration graph for the differential pressure sensor.

3. Results and Discussion

A standard representative 14 mm automotive spark plug (Champion RS12YC) was used for all of the tests. The center electrode (cathode) had a diameter of 2.5 mm and the width of the J-type ground strap was also 2.5 mm. It had a nominal internal resistance of 6.1 kohms. The inductive-type ignition coil had a secondary inductance value of 30 H. The default dwell time used was 4 ms. However, the effect of dwell times of 2 and 6 ms were studied as well. Figure 5a presents an example of the measured current delivered to the spark plug and a voltage trace corrected for the voltage drop across the internal resistor, and thus, represents the gap voltage. The integrated product of the corrected gap voltage and the current was used to calculate the electrical energy delivered to the gap. These representative traces were for conditions of 8 bar absolute gas pressure, with a 0.9 mm gap and a dwell time of 4 ms. Figure 5b shows a typical voltage trace having a short time scale as used to measure the breakdown voltage. The breakdown voltage for this trial was 8.67 kV and the spark duration was 3.45 ms.

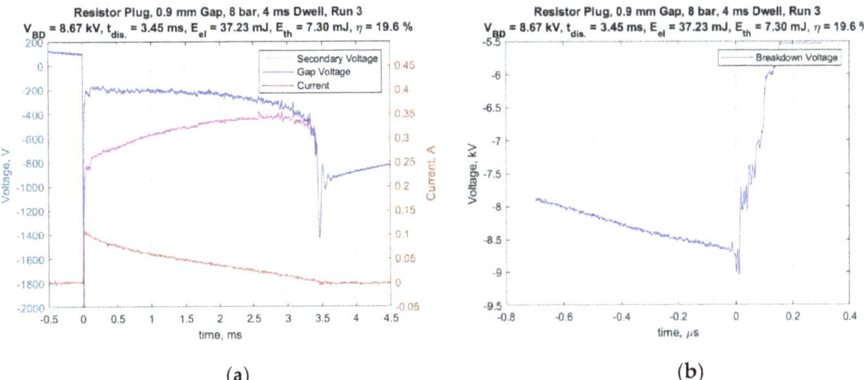

Figure 5. (**a**) Current and voltage vs. time for entire discharge duration and (**b**) voltage vs. time for breakdown.

It is interesting to note that for all gas pressures above 1 atm, the ignition system produced an arc-type discharge, whereas for 1 atm pressure and the smallest spark gap of 0.3 mm, an arc to glow transition occurred immediately following breakdown. This was readily distinguished by the level of the relatively flat voltage profile following breakdown. The arc discharges were in a range of approximately 100–300 volts, whereas glow discharges were in the range of approximately 400–500 volts. For all of the measurements, the peak current following breakdown was approximately 100 mA and decreased nearly linearly with time, but with different spark durations.

Figure 6a shows an example of recorded pressure traces for different initial gas pressures ranging from 1 to 24 bar. In general, the pressure rise was very rapid during the early part of the discharge, consistent with the high current levels during that period and representative of the large energy deposition during the breakdown process and shortly thereafter.

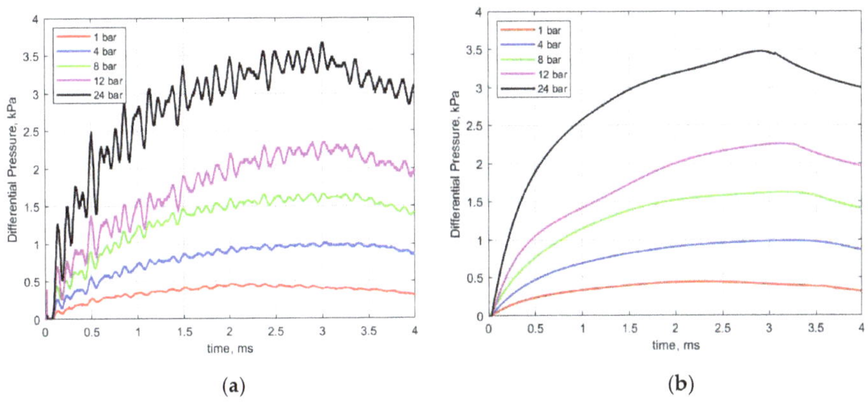

Figure 6. (**a**) Recorded pressure rise vs. time for different initial gas pressures and (**b**) smoothed pressure traces used for analysis. Gap distance 0.9 mm; dwell 4 ms.

As seen in Figure 6a, there is a periodic ringing superimposed upon the pressure signal. For the 1 atmosphere case, this sinusoidal pressure variation superimposed on the signal has a frequency of approximately 2.4 kHz. Considerable effort was made to understand the cause of this. It was initially attributed to an acoustic mode excited in the short (4 mm in length) passage tube that was part of the pressure sensor. To test for this, the passage tube was cut off and a new holder for the pressure sensor was machined to fit and seal the new sensor geometry. Subsequent tests revealed that the ringing remained but that it occurred at an increased frequency of approximately 6.1 kHz. At this point, a more detailed acoustic analysis was performed that considered possible acoustic modes associated with the calorimeter chamber and also considered the system response of the pressure transducer.

The calorimeter chamber is to first order a right circular cylinder with rigid acoustic boundary conditions and is the largest acoustic cavity in the system, and hence must exhibit the lowest acoustic cavity resonance frequency. Given the aforementioned dimensions, and considering the 1 atmosphere case and room temperature, the lowest-order resonance frequency is for approximately one-half wavelength acoustic fluctuations standing in the largest spatial dimension (12.8 mm), which yields a resonance frequency of approximately 13.4 kHz. All other acoustic standing wave modes must be higher in frequency. For example, the next largest dimension is the length of the calorimeter chamber, which at 9.5 mm yields approximately 18 kHz for the lowest-order resonance frequency. These frequencies are significantly higher than the fluctuations observed in the experimental data, and hence the observed ringing should not be attributed to acoustic standing wave modes in the system. To verify this, a spark plug and the pressure sensor were placed close together outside of the calorimeter, and operated in free space. A spark discharge was measured and the ringing in the pressure signal was still observed.

Below the lowest-order standing wave modal frequency, acoustic lumped-parameter behavior can occur and the pressure sensor contains elements that form a Helmholtz resonator. The air inside the passage tube acts as an acoustic mass, and the air inside the cavity within the sensor acts as an acoustic compliance. Together, these form a simple harmonic oscillator, which can be excited by a broadband source of energy, such as the shock generated by the spark [15]. Examination of both the sensor's specification sheet and the sensor itself yielded estimates of the dimensions of the structures required for the calculation of the Helmholtz resonance frequencies using

$$f_0 = \frac{c}{2\pi}\left(\frac{S}{L'V}\right)^{\frac{1}{2}} \tag{1}$$

where $c = 343$ m/s is the speed of sound at room conditions, S is the inner area of the passage tube, and V is the volume of the cavity. The effective length of the passage tube is $L' = L + 1.4a$, where L is the actual length of the passage tube and a is its inner radius [16]. For the case with the passage tube attached, the predicted Helmholtz resonance frequency is 2.4 kHz and the observed frequency was 2.1 kHz. With the passage tube removed, $L = 0$, and the appropriate effective length is $L' = 1.7a$ [16]. The predicted Helmholtz resonance frequency is 6.4 kHz and the observed frequency was 6.1 kHz. Both observations match the predictions within the uncertainty of the estimate of the dimensions, hence the observed fluctuations appear to be an artifact due to Helmholtz resonator behavior of the sensor.

The tabulated sensor design specifications include a first-order system time constant of 0.45 ms, which in turn yields a low-pass system response with an upper band limit of approximately 350 Hz, hence the sensor is not designed to accurately measure the effects of the initial shock, nor dynamic pressure fluctuations above approximately 350 Hz. Before performing the analyses relating the pressure rise to the thermal energy deposition, the pressure curves were smoothed to remove the Helmholtz resonator artifacts, as shown in Figure 6b.

The thermal energy deposition to the chamber gas was derived from the measured pressure rise using Equation (2) [13]. Of the electrical energy delivered to the gap, a portion is retained as thermal energy within the gas with the remainder lost as heat transfer, primarily to the electrodes. Equation (2) expresses the effect of the retained thermal energy of the gas on the rise in pressure assuming chemical and thermal equilibrium and using the simplifying assumptions of constant specific heat and ideal gas behavior.

$$E_{therm} = \frac{V}{\gamma - 1}\Delta P \tag{2}$$

In Equation (2), V is the chamber volume, ΔP is the maximum pressure rise and γ is the ratio of specific heats of nitrogen. The relative contributions of possible heat loss mechanisms are not clear.

The difference in temperature between the plasma/gas and the spark plug surfaces will contribute to thermal heat transfer, while there may also be Ohmic losses along the surfaces of the electrodes. To be representative of an engine, the heat losses should reflect the sources of loss in an engine, and not for example, due to heat loss to the walls of the calorimeter. A small decrease in chamber pressure was sometimes observed toward the end of the spark discharge process (e.g., Figure 6) indicating continued heat transfer. It is likely that this thermal heat transfer was to the spark plug electrodes and not to the chamber walls over this period since an estimate of molecular thermal diffusion to the chamber walls yields a characteristic time of 100 s of milliseconds. In this study, we defined the amount of thermal energy delivered to the gas to be that reflected by the maximum of the measured pressure rise. Over the short duration of the spark, the gas is not in thermal equilibrium, as temperature gradients exist. As a result, the application of Equation (2) implies a spatial mass average of the thermal contributions to the pressure of the different temperature layers. This should be valid because of the linear relationship between mass, temperature and the pressure as given by the ideal gas law and the linear relationship between thermal energy and temperature inherent in the constant specific heat assumption. In addition, during the spark discharge, not all of the gas exists as N_2 since there is dissociation of the nitrogen in the spark plasma; however, only a small fraction of the gas is so affected and equilibrium processes will have largely returned this gas to N_2 by the time the spark event has ended.

The measured spark breakdown voltages as functions of the initial gas pressure and spark gap are shown in Figures 7 and 8 for a dwell time of 4 ms. Each data point represents an average of five measurements.

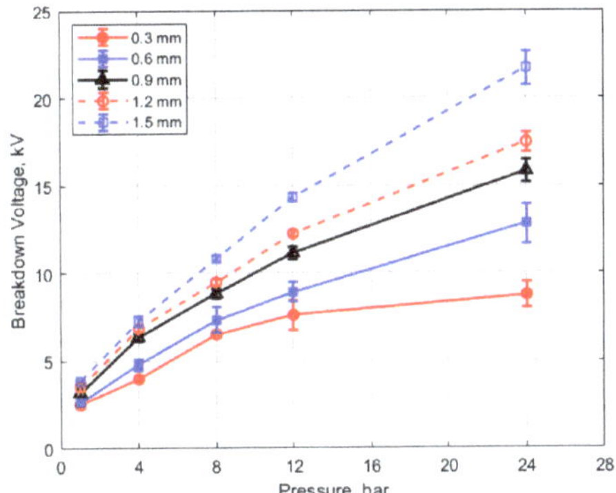

Figure 7. Breakdown voltage vs. initial pressure for different gaps, for a dwell time of 4 ms.

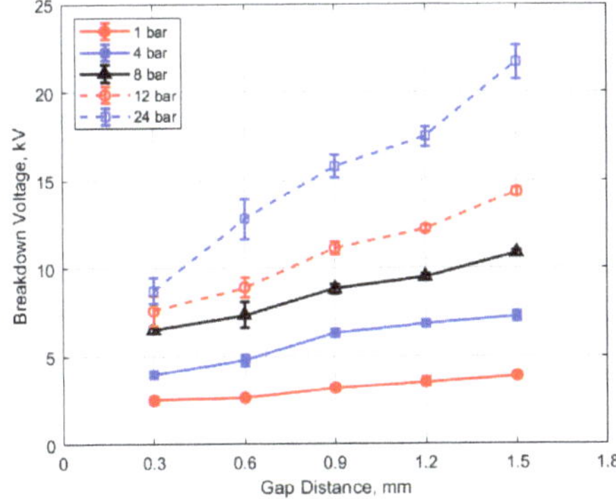

Figure 8. Breakdown voltage vs. gap for different pressures, for a dwell time of 4 ms.

As expected, the breakdown voltage increases with both the gas pressure and the gap distance, but it is interesting to note that the breakdown voltage was not directly proportional to either of these variables even though linear relationships are predicted by Paschen's law [5]. In the case of gas pressure, the breakdown voltage increased at a rate that diminished as the gas pressure increased. While the trend was closer to a linear increase for a changing spark gap, there was a considerable zero offset for the breakdown voltage at all of the pressures for the data extrapolated back to zero gap. In Figures 9 and 10, the of effect dwell time on breakdown voltage is shown. There was no appreciable effect of either gap distance or gas pressure on the breakdown voltage within the uncertainty of the measurements over the dwell times ranging from 2 to 6 ms; however, a very slight increase in breakdown voltage with increasing dwell time is suggested from Figure 9. Since a longer dwell time results in more energy stored in the secondary coil and since the energy stored in an inductor follows Equation (3),

$$E = \frac{1}{2}LI^2 \tag{3}$$

where L is the coil inductance and I is the current, it follows that the current will be greater in the inductor at higher energies. It may be the case that for the higher secondary currents associated with greater stored energies, the electric field in the gap rises slightly faster prior to breakdown, resulting in a higher voltage at the gap before the actual breakdown occurs.

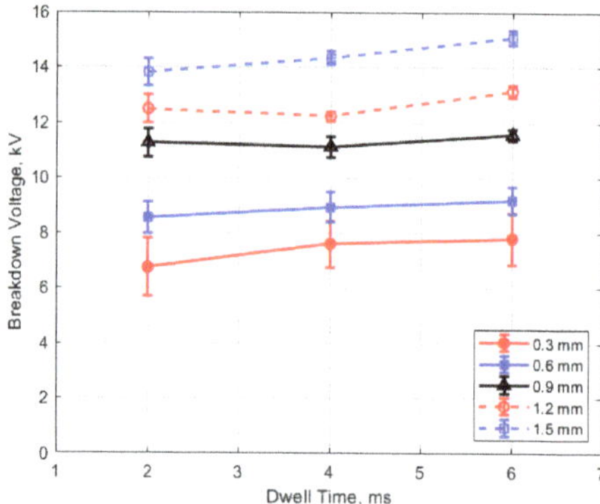

Figure 9. Breakdown voltage vs. dwell for different gaps, for a pressure of 12 bar.

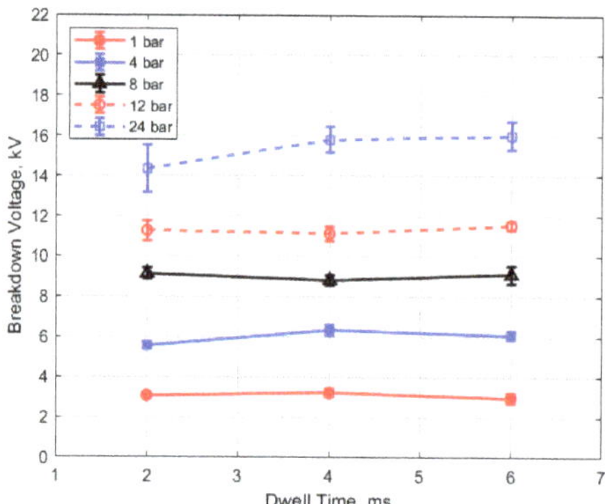

Figure 10. Breakdown voltage vs. dwell for different pressures, for a 0.9 mm gap.

The discharge time or spark duration as a function of gas pressure, gap distance, and dwell time is given in Figures 11–14. As seen in Figure 11, the spark duration is a weak function of gas pressure, but generally decreases with increasing pressure. The exceptions to this were the measurements at

1 atm, which show a shorter spark duration. This was associated with the glow-type discharge that was mentioned previously. The discharge times were in the range of approximately 3–4 ms. Figure 12 shows that the discharge times tended to shorten slightly as the gap distance increased; however, there is an exception for the 1 atm measurements. Figures 13 and 14 show that spark duration increased with dwell time as more electrical energy was delivered to the gap, with the increase weaker from 4 to 6 ms than from 2 to 4 ms.

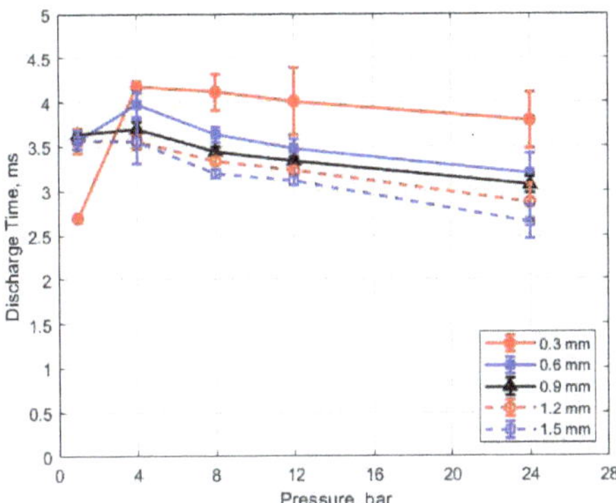

Figure 11. Discharge time vs. pressure for different gaps, for a dwell time of 4 ms.

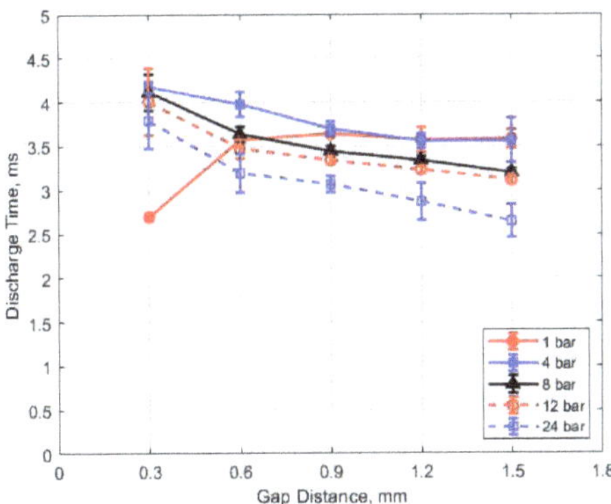

Figure 12. Discharge time vs. gap for different pressures, for a dwell time of 4 ms.

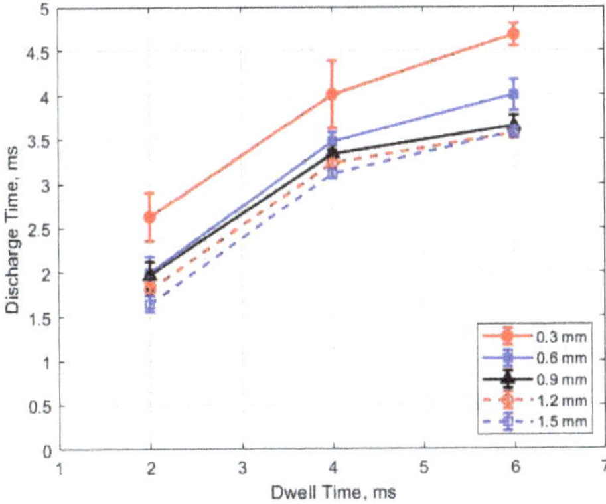

Figure 13. Discharge time vs. dwell for different gaps, for a pressure of 12 bar.

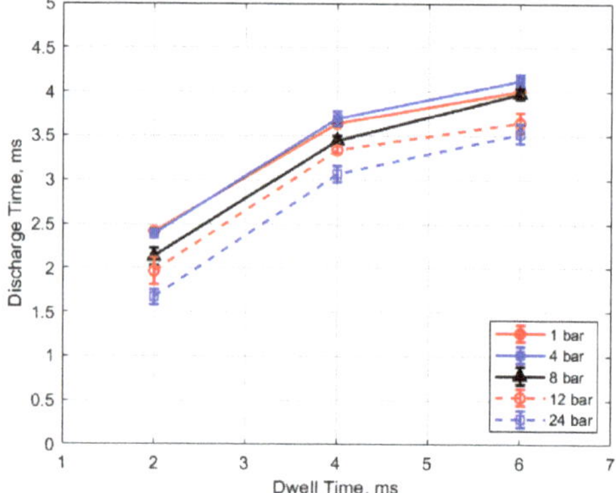

Figure 14. Discharge time vs. dwell for different pressures, for a 0.9 mm gap.

Figures 15–18 show the calculated electrical energy delivered to the gap based on the measured current traces and the corrected traces of gap voltage. From Figures 15 and 16, it can be seen that the electrical energy delivered to the gap generally increased, both with increasing gas pressure and with gap distance. As both the gas pressure and gap distance increase, the breakdown voltage increases as well; increasing the peak voltage in the secondary side of the ignition circuit will lead to greater electrical energy storage according to the secondary circuit and spark plug capacitance (typically on the order of 15 pF [10]). Both higher gas pressure and larger gap result in a greater gap resistance and in a more thermally isolated gap region since molecular diffusivities scale as the square of the characteristic diffusion distance. It is interesting that electrical energy delivered to the gap increased as these two parameters increased even though spark duration decreased. The relatively high electrical energy delivery for the case of 1 atm gas pressure and the smallest gap of 0.3 mm was a consequence of the high glow voltage relative to the arc voltages; the current traces were similar for arc and glow.

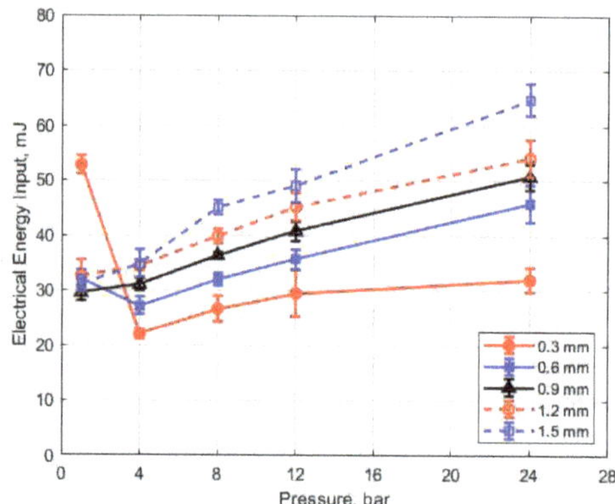

Figure 15. Electrical energy input vs. pressure for different gaps, for a dwell time of 4 ms.

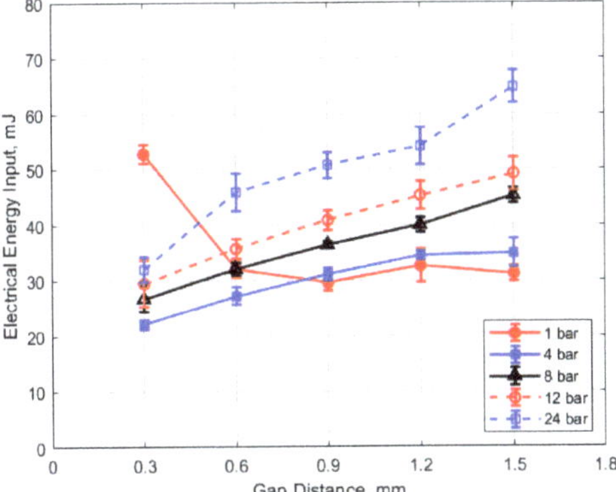

Figure 16. Electrical energy input vs. gap for different pressures, for a dwell time of 4 ms.

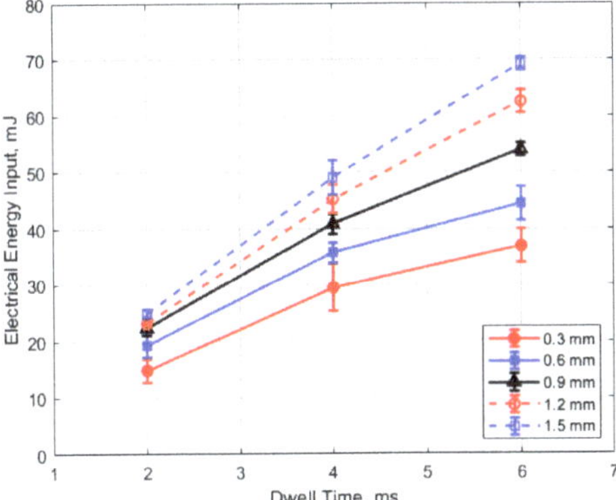

Figure 17. Electrical energy input vs. dwell for different gaps, for a pressure of 12 bar.

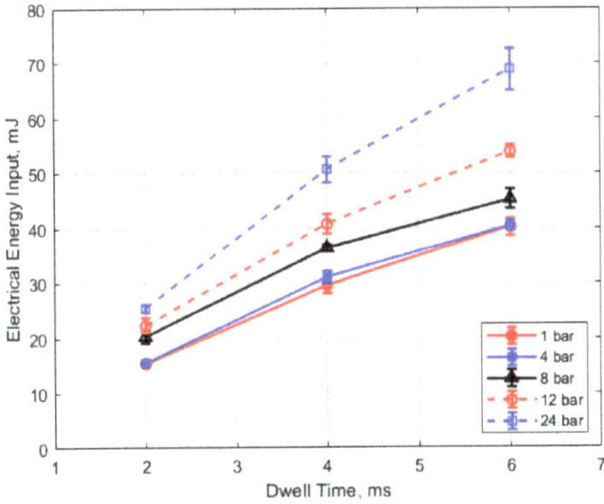

Figure 18. Electrical energy input vs. dwell for different pressures, for a 0.9 mm gap.

The electrical energy delivered to the gap as a function of dwell time is shown in Figures 17 and 18. As expected, a longer dwell time leads to greater electrical energy storage in the secondary side of the coil and, consequently, more energy delivered to the gap. It is interesting to note the range of delivered electrical energy to the gap as parameters changed; delivered energies ranged from approximately 15 to 70 mJ. The literature shows minimum ignition energies for stoichiometric hydrocarbon-air mixtures that are in the range of 0.3–2 mJ [17].

The thermal energy deposition to the gas is depicted in Figures 19–22. The quantity of thermal energy delivered to the gas and its distribution in the vicinity of the spark gap is important for flame initiation, particularly for conditions for which the development of a strong flame kernel is challenging, such as in lean, highly dilute and inhomogeneous mixtures and where fluid motion near the gap can strongly convect both the spark plasma and the developing flame kernel.

Figure 19 shows the thermal energy deposition versus gas pressure as a function of gap distance and for a dwell time of 4 ms. The thermal energy delivered to the gas is seen to increase monotonically with increasing pressure. If one considers the results for the different gap distances, the normalized change in thermal energy deposition with pressure is similar for all of them; that is

$$\frac{E(P_2) - E(P_1)}{P_2 - P_1} \approx \text{constant} \tag{4}$$

where E is the thermal energy to the gas in mJ and P is the gas pressure in bar. For the pressures shown, the value of the constant in Equation (4) is in the range of 0.2–0.4 mJ/bar, with an average close to 0.3 mJ/bar. Using this average value of 0.3 mJ/bar and rearranging Equation (4) as a differential, one obtains the following relationship for the differential change in the normalized thermal energy deposition with gas pressure that is independent of gap distance.

$$\frac{dE}{E} = \left(0.3\text{bar}^{-1}dP\right) - 1 \tag{5}$$

It would require more investigation to determine the validity of Equation (5) for different spark plug geometries and internal resistances, the type of ignition circuit, dwell times, and of course, gap fluid motion, but it may provide an approximate scaling rule in the absence of other data. It may be more appropriate to express this relationship in terms of gas density rather than pressure through their relationship via the ideal gas law.

Figure 20 shows the thermal energy deposition versus gap distance as a function of initial absolute gas pressure. Thermal energy deposition increased monotonically with gap distance for the entire range of gas pressures. Differences in gap electrical resistance and differences in heat transfer to the electrodes are probably responsible for these trends. As the gap distance increases, the positive column portion of the spark plasma increases proportionally, along with its electrical resistance. Molecular thermal diffusivities increase as the square of the characteristic heat transfer length, which one might take as one-half the gap distance. Despite the possible non-linearity, it is interesting to note that increases in thermal energy deposition with gap distance follow an approximately linear trend.

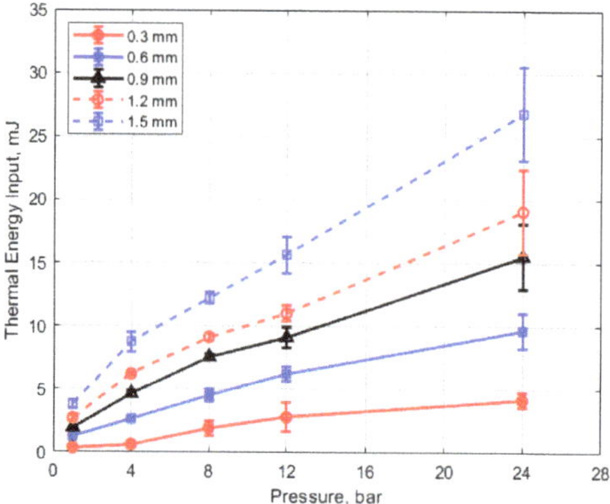

Figure 19. Thermal energy input vs. pressure for different gaps, for a dwell time of 4 ms.

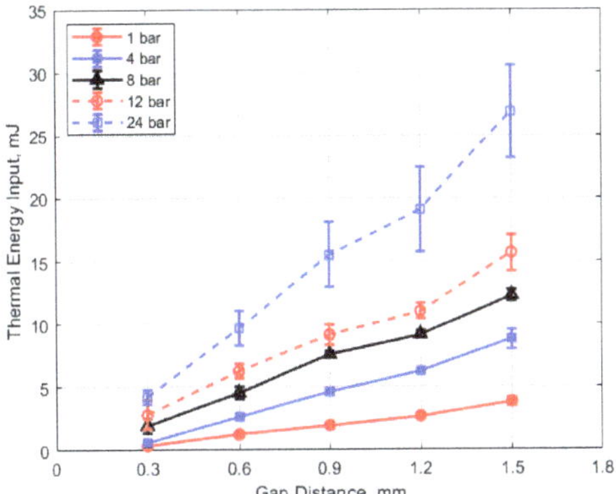

Figure 20. Thermal energy input vs. gap for different pressures, for a dwell time of 4 ms.

An analysis for the dependence of the change in deposition energy with gap distance, analogous to that carried out for the pressure dependence, can be performed. The related constant from the experimental data is in the range of approximately 4–6 mm^{-1}, with an average value close to 5 mm^{-1}. Equation (6) is the resulting relationship for the effect of gap distance "g" on the change in thermal energy deposition that is independent of gas pressure

$$\frac{dE}{E} = \left(5\text{mm}^{-1}dg\right) - 1 \tag{6}$$

Again, the sensitivity to other parameters needs to be explored further.

Figures 21 and 22 show the thermal energy deposition as a function of dwell. A longer dwell time results in more electrical energy delivered to the gap, so the observed trend of increasing thermal energy deposition with increasing dwell time would be expected.

Finally, the conversion efficiencies of electrical energy to thermal energy delivered to the gap are shown in Figures 23–26. Figure 23 shows the conversion efficiency as a function of initial gas pressure for the different gap distances. Conversion efficiency increased monotonically with gas pressure, but the dependency was not linear.

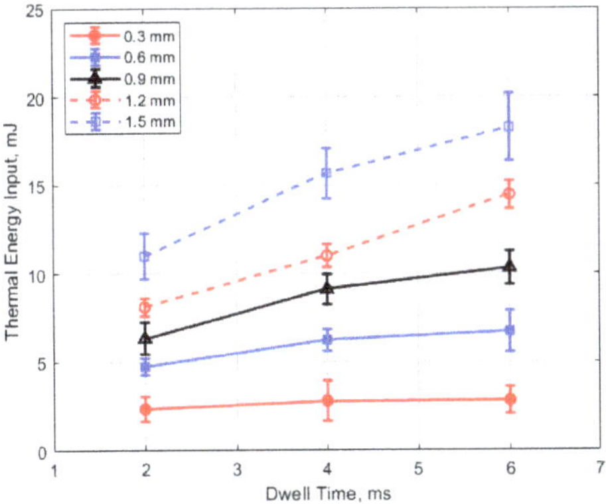

Figure 21. Thermal energy input vs. dwell for different gaps, for a pressure of 12 bar.

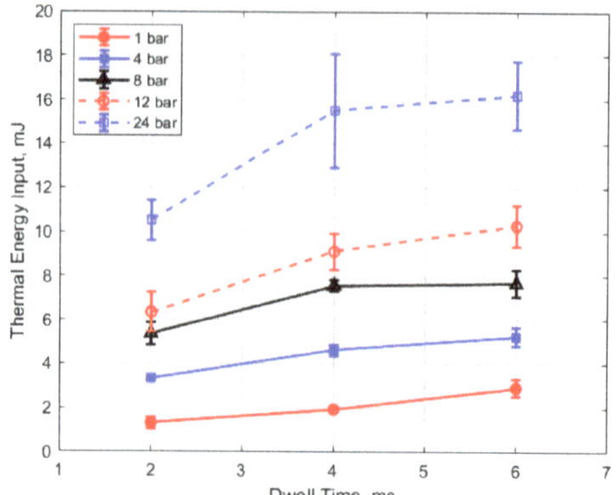

Figure 22. Thermal energy input vs. dwell for different pressures, for a 0.9 mm gap.

The degree to which the conversion efficiency increased with gas pressure was quite striking. At a pressure of 1 atm and for the smallest gap distance of 0.3 mm, the conversion efficiency was less than 1%, whereas with the largest gap of 1.5 mm and at the highest gas pressure of 24 bar, the conversion efficiency was over 40%. The rate of increase in the conversion efficiency with pressure tended to diminish with increasing pressure, in contrast to the thermal energy deposition for which the rate of increase showed an increase with increasing gas pressure.

Of course, the gas density associated with the room temperature measurements at 1 atm are not representative of the gas density in an engine at the time of ignition. However, the laboratory gas densities at pressures starting at 4 bar become relevant to engine ignition. At 4 bar pressure, conversion efficiencies were less than 15% for gap distances less than 1 mm.

The dependence of conversion efficiency on spark gap distance was also very strong, as depicted in Figure 24. A smaller gap distance effectively increases the ratio of the cold metal surface area subject to heat transfer relative to the volume of the spark plasma, leading to greater heat losses from the arc.

The conversion efficiencies versus dwell time are shown in Figures 25 and 26. A long dwell time results in greater electrical energy supplied to the gap. However, it is interesting that, in general, the efficiency of electrical-to-thermal energy conversion diminished with increasing dwell time. The reason for this is not clear, but longer dwell time resulted in longer spark durations and, consequently, more time for heat transfer to occur.

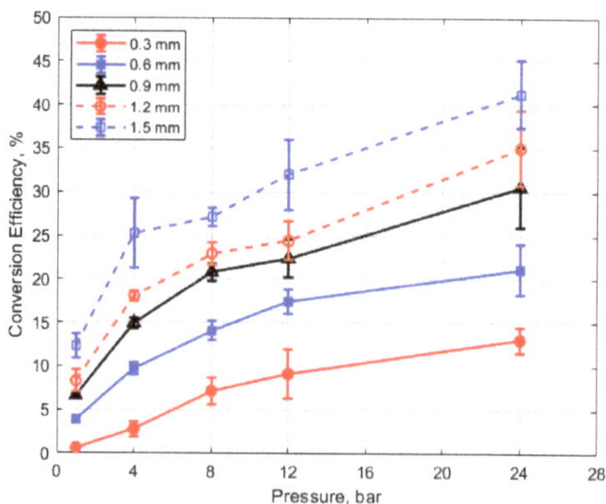

Figure 23. Conversion efficiency vs. pressure for different gaps, for a dwell time of 4 ms.

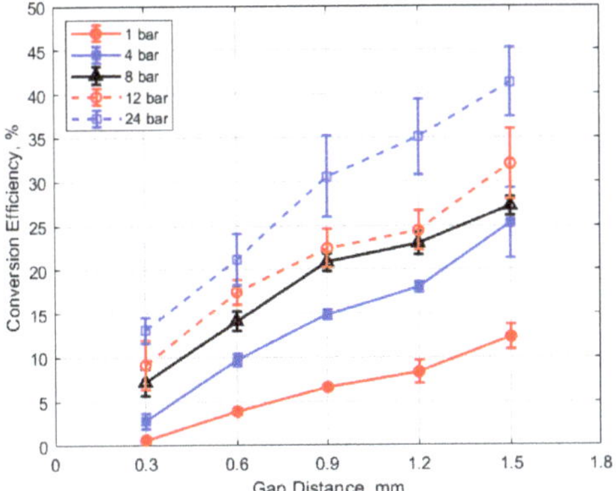

Figure 24. Conversion efficiency vs. gap for different pressures, for a dwell time of 4 ms.

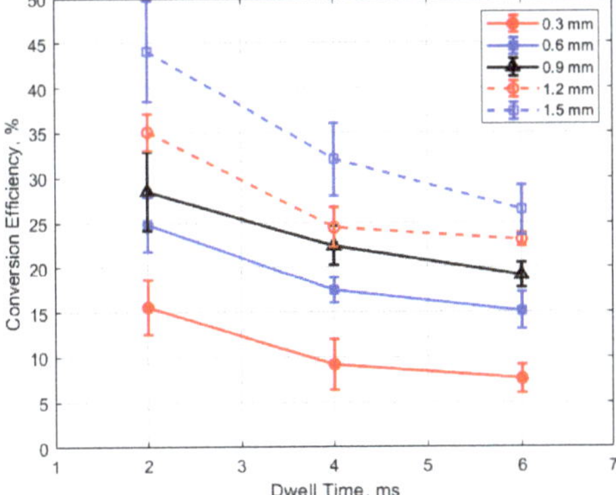

Figure 25. Conversion efficiency vs. dwell for different gaps, for a pressure of 12 bar.

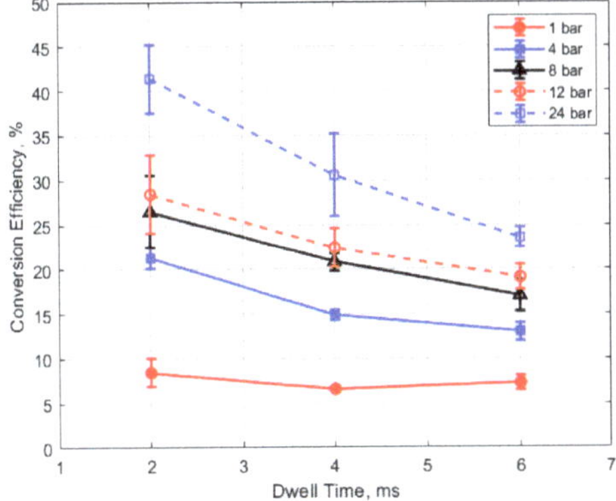

Figure 26. Conversion efficiency vs. dwell for different pressures, for a 0.9 mm gap.

4. Summary and Conclusions

A spark plug calorimeter was designed and built that utilizes a high sensitively chip-based differential pressure sensor to detect the rise in gas pressure associated with the thermal energy deposition to the gas from the spark discharge. It differs from most spark plug calorimeters in that the differential pressure rise is measured relative to the initial pressure of the calorimeter chamber rather than relative to the outside ambient pressure or internal reference. This feature creates the possibility of higher precision measurements from a wider dynamic range of the sensor and its high sensitivity. The strength of its stainless steel construction allowed tests to be safely performed at gas pressures at least as high as 24 bar.

The present calorimeter was used to measure the thermal energy deposition and the efficiency of electrical energy to thermal energy conversion for arc discharges for a standard 14 mm automotive resistor-type spark plug for gas pressures between 1 atm and 24 bar, gap distances between 0.3 and 1.5 mm, and inductive ignition circuit dwell times between 2 and 6 ms.

Spark breakdown voltages were measured as well. They were in the expected range, based on the literature, but they did not closely follow the expected linear trends relative to changes in gas density and gap distance suggested by Paschen's law (e.g., Figures 7 and 8), with the rate of increase in breakdown voltages with increasing gap and increasing pressure less than linear.

Arc-type, as opposed to glow-type spark discharges, were investigated, as distinguished by the sustaining arc voltages during discharge in the range of 100 V–200 V (e.g., Figure 5). The measured electrical energies delivered to the gap increased with both increasing gas pressure and gap distance, but the thermal energy delivered to the gap increased much more strongly. For example, delivered electrical energies, for the larger gaps, increased by approximately 50–90% between 4 and 24 bar pressures, whereas thermal energy deposition increased approximately 200% over the same pressure range. For the same ignition hardware and operating conditions, the thermal energy delivered to the gap varied from less than 1 mJ at 1 atm pressure and a gap distance of 0.3 mm to over 25 mJ at a pressure of 24 bar and a gap distance of 1.5 mm (e.g., Figures 19 and 20).

These trends were reflected in the electrical-to-thermal energy conversion efficiency, which increased monotonically with gas density. For gas densities representative of those in an engine at the time of ignition, the conversion efficiencies ranged from approximately 3% at low pressures (4 bar) and small gap (0.3 mm) to as much as 40% at the highest pressure of 24 bar and with a gap of 1.5 mm (e.g., Figures 23 and 24).

For the particular spark plug and ignition circuit used, empirically calibrated expressions were derived for the incremental change in thermal energy deposition with changes in either gas pressure or gap distance.

Author Contributions: K.K. was responsible for data curation, validation and formal analysis. P.S.W. was responsible for the acoustic analysis of the calorimeter/sensor system. R.D.M. and M.J.H. were responsible for the conceptualization and supervision of the project. All authors participated in the writing, review, and editing of the manuscript. All authors have read and agreed to the published version of the manuscript.

Funding: This project was made possible through the funding provided by Cummins Inc, through the University of Texas at Austin's site of the National Science Foundation Center for Efficient Vehicles and Sustainable Transportation Systems (EVSTS), NSF Award 1650483.

Acknowledgments: The authors wish to express their gratitude to Sachin Joshi, Daniel J. O'Connor and Douglas L. Sprunger of Cummins Inc. for many helpful discussions and we wish to thank Cummins Inc. for their financial support of this project.

Conflicts of Interest: The authors declare no conflicts of interest.

References

1. Maly, R.; Vogel, M. Initiation and Propagation of Flame Fronts in Lean CH4-Air Mixtures by the Three Modes of the Ignition Spark. In Proceedings of the 17th Symposium on Combustion, the Combustion Institute, Leeds, UK, 20–25 August 1978; pp. 821–831.
2. Haiwen, G.; Zhao, P. A Comprehensive Ignition System Model for Spark Ignition Engines. In Proceedings of the ASME 2018, Internal Combustion Engine Division Fall Technical Conference, ICEF2018-9574, San Diego, CA, USA, 4–7 November 2018.

3. Scarcelli, R.; Zhang, A.; Wallner, T.; Som, S.; Huang, J.; Wijeyakulasuriya, S.; Mao, Y.; Zhu, X.; Lee, S. Development of a Hybrid Lagrangian-Eularian Model to describe Spark-Ignition Processes at Engine-like Turbulent Flow Conditions. In Proceedings of the ASME 2018, Internal Combustion Engine Division Fall Technical Conference, ICEF2018-9690, San Diego, CA, USA, 4–7 November 2018.
4. Subramaniam, V.; Karpatne, A.; Breden, D.; Laximinarayan, R. Simulation of Spark-Initiated Combustion. In Proceedings of the WCX SAE World Congress Experience, Detroit, MI, USA, 9–11 April 2019; SAE Technical Paper 2019-01-0226.
5. Meek, N.F. *Electrical Breakdown of Gases*; Oxford at the Clarendon Press: London, UK, 1953.
6. Maly, R. Spark Ignition: Its Physics and Effect on the Internal Combustion Engine. In *Fuel Economy*; Hilliard and Springer, Ed.; Chapter 3; Plenum Press: New York, NY, USA, 1984.
7. Lee, M.; Hall, M.J.; Ezekoye, O.A.; Matthews, R.D. Voltage and Energy Deposition Characteristics of Spark Ignition Systems. In Proceedings of the SAE 2005 World Congress & Exhibition, Detroit, MI, USA, 11–14 April 2005; SAE Technical Paper 2005-01-0231.
8. Roth, W.; Guest, P.G.; Elbe, G.; Lewis, B. Heat Generation by Electric Sparks and Rate of Heat Loss to the Spark Electrodes. *J. Chem. Phys.* **1951**, *19*, 1530–1535. [CrossRef]
9. Merritt, L.R. A Spark Calorimeter. *J. Phys. E Sci. Instrum.* **1978**, *11*, 193–194. [CrossRef]
10. Franke, A.; Reinmann, R. Calorimetric Characterization of Commercial Ignition Systems. In Proceedings of the SAE 2000 Automotive Dynamics & Stability Conference SAE 2000 World Congress, Troy, MI, USA, 15–17 May 2000; SAE Technical Paper 2000-01-0548. [CrossRef]
11. Teets, R.E.; Sell, J.A. Calorimetry of ignition sparks. In Proceedings of the SAE International Congress and Exposition, Detroit, MI, USA, 29 February 1988; SAE Technical Paper 880204.
12. Verhoeven, D. Spark heat transfer measurements in flowing gases. In Proceedings of the 1995 SAE International Fall Fuels and Lubricants Meeting and Exhibition, Toronto, ON, Canada, 16–18 October 1995; SAE Technical Paper 952450.
13. Abidin, Z.; Chadwell, C. Parametric Study and Secondary Circuit Model Calibration Using Spark Calorimeter Testing. In Proceedings of the SAE 2015 World Congress & Exhibition, Detroit, MI, USA, 21–23 April 2015; SAE Technical Paper 2015-01-0778. [CrossRef]
14. Alger, T.; Mangold, B.; Mehta, D.; Roberts, C. The Effect of Sparkplug Design on Initial Flame Kernel Development and Sparkplug Performance. In Proceedings of the SAE 2006 World Congress & Exhibition, Busan, Korea, 24–28 September 2006; SAE Technical Paper 2006-01-0224. [CrossRef]
15. Ekici, O.; Ezekoye, O.A.; Hall, M.J.; Matthews, R.D. Thermal and Flow Fields Modeling of Fast Spark Discharges in Air. *ASME J. Fluids Eng.* **2007**, *129*, 55–65. [CrossRef]
16. Kinsler, L.E.; Frey, A.R. *Fundamentals of Acoustics*; Wiley: New York, NY, USA, 2000.
17. Lewis, B.; von Elbe, G. *Combustion, Flames, Explosions of Gases*; Academic Press: New York, NY, USA, 1987.

Article

Experimental Study of Premixed Gasoline Surrogates Burning Velocities in a Spherical Combustion Bomb at Engine Like Conditions

Miriam Reyes *, Francisco V. Tinaut and Alexandra Camaño

Department of Energy and Fluid-Mechanics Engineering, University of Valladolid, Paseo del Cauce, 59, 47011 Valladolid, Spain; tinaut@eii.uva.es (F.V.T.); alexandra.lisbeth@gmail.com (A.C.)

* Correspondence: miriam.reyes@uva.es

Received: 10 June 2020; Accepted: 1 July 2020; Published: 3 July 2020

Abstract: In this work are presented experimental values of the burning velocity of iso-octane/air, n-heptane/air and n-heptane/toluene/air mixtures, gasoline surrogates valid over a range of pressures and temperatures similar to those obtained in internal combustion engines. The present work is based on a method to determine the burning velocities of liquid fuels in a spherical constant volume combustion bomb, in which the initial conditions of pressure, temperature and fuel/air equivalence ratios can be accurately established. A two-zone thermodynamic diagnostic model was used to analyze the combustion pressure trace and calculate thermodynamic variables that cannot be directly measured: the burning velocity and mass burning rate. This experimental facility has been used and validated before for the determination of the burning velocity of gaseous fuels and it is validated in this work for liquid fuels. The values obtained for the burning velocity are expressed as power laws of the pressure, temperature and equivalence ratio. Iso-octane, n-heptane and mixtures of n-heptane/toluene have been used as surrogates, with toluene accounting for the aromatic part of the fuel. Initially, the method is validated for liquid fuels by determining the burning velocity of iso-octane and then comparing the results with those corresponding in the literature. Following, the burning velocity of n-heptane and a blend of 50% n-heptane and 50% toluene are determined. Results of the burning velocities of iso-octane have been obtained for pressures between 0.1 and 0.5 MPa and temperatures between 360 and 450 K, for n-heptane 0.1–1.2 MPa and 370–650 K, and for the mixture of 50% n-heptane/50% toluene 0.2–1.0 MPa and 360–700 K. The power law correlations obtained with the results for the three different fuels show a positive dependence with the initial temperature and the equivalence ratio, and an inverse dependence with the initial pressure. Finally, the comparison of the burning velocity results of iso-octane and n-heptane with those obtained in the literature show a good agreement, validating the method used. Analytical expressions of burning velocity as power laws of pressure and unburned temperature are presented for each fuel and equivalence ratio.

Keywords: iso-octane; n-heptane; toluene; surrogate fuels; burning velocity; combustion bomb

1. Introduction

The laminar burning velocity is a fuel property of fundamental importance for predicting and studying the performance of internal combustion engines, which can be extremely useful in the analysis of fundamental processes and serve as a design utility during the engine design stage. The laminar burning velocity of premixed flames has been the focus of comprehensively experimental and numerical investigations. This property is essential in the analysis, design and implementation of internal combustion engines (ICE), because the burning velocity has a direct influence on the efficiency, emissions and burn rate in the combustion engine [1]. Burning velocities are implemented in combustion models to validate kinetic mechanisms. The laminar burning velocity of fuel mixtures can be obtained using several techniques, typically by combining experimental methods (pressure register, optical) with thermodynamic models and instabilities studies to consider the effects of flame stretch.

Precise measurements of values of the burning velocity are necessary to characterize fuels in premixed combustions and to validate combustion models [2]. A technique widely used in the literature, to present the results of the combustion rates of different fuels, is a correlation (power law) based on the initial conditions of the pressure, temperature and equivalence ratio of a given fuel mixture for a determined range of the pressure and temperature. These correlations provide very important information for evaluating the effect of fuels in spark ignition (SI) engines.

Due to the complexity of the gasoline composition and its variability, gasoline surrogates are used both experimentally and numerically to simplify calculations. Gasoline surrogates refers to fuels with a simpler representation of a fully mixed fuel, which can be primary reference fuels or binary mixtures based on the research octane number. Some of these gasoline surrogates are used as fuels for advanced combustion engines [3]. Surrogate fuels have a simpler composition which can facilitate simulations and evaluate the property effects and fuel-composition of the in-cylinder processes: vaporization, mixing and combustion, determining the processes in engine efficiency, emissions, performance, and requirements about the after-treatment systems [4]. Therefore, surrogate fuels have a great value as reference fuels which can be used to determine different parameters of interest in engine combustion without the effect of changes in the fuel-composition. The most common gasoline surrogates are iso-octane and n-heptane, the primary reference fuels (usually named PRF's) or binary mixtures of them for determining the research octane numbers (RON) in SI fuels.

Iso-octane and n-heptane are standard gasoline surrogates for modeling combustion in SI engines because the processes of oxidation of n-heptane and iso-octane represent the ignition process and combustion characteristics of a gasoline fuel. For that reason, the effect of composition changes of multicomponent fuels can be simulated using binary mixtures of them as a first approximation [5].

Some investigations have been developed to determine the laminar burning velocity of iso-octane, n-heptane and binary mixtures of these two fuels, and, for that reason, iso-octane and n-heptane are used in this work to validate the methodology presented. Gülder [6], Bradley et al. [7], Metghalchi et al. [8], Galmiche et al. [9] and Müller et al. [10] obtained different correlations for the burning velocity of iso-octane as a function of pressure and temperature for diverse fuel/air equivalence ratios. Galmiche et al. [9] measured flame velocities in spherically expanding flames, from which the corresponding laminar burning velocities at a stretch rate of zero are derived. Iso-octane/air mixtures at initial temperatures between 323 and 473 K, and pressures between 0.1 and 1.0 MPa are studied over an extensive range of equivalence ratios, using a high-speed shadowgraph system. Varea et al. [11] used a high pressure and temperature combustion chamber to obtain the laminar burning velocity of iso-octane with a new method, from the difference between the flame speed and the intake gas velocity. They also obtained correlations as a function of pressure and temperature to express their effect. Marshal et al. [12] obtained the laminar burning velocity of iso-octane and n-heptane at elevated pressures and temperatures and they included combustion residuals in their experiments. They used a constant volume combustion vessel equipped with a Schlieren technique and obtained values for different ranges of pressure, temperature and equivalence ratios. The burning velocity of n-heptane has been obtained by different researchers: Davis and Law (1998, [13]) used a counterflow twin flame configuration, Huang et al. (2004, [14]) obtained the burning velocity of n-heptane and iso-octane (PRFs) in a counterflow configuration using a digital particle image velocimetry, Van Lipzig et al. (2011, [15]) investigated the adiabatic laminar burning velocity of n-heptane and iso-octane and their mixtures in a perforated plate burner, Sileghem et al. (2013, [16]) used the heat flux method to obtain the laminar burning velocities of iso-octane, n-heptane and toluene, Kwon et al. (2000, [17]) measured burning velocities in a spherical windowed chamber, Kumar et al. (2007, [18]) obtained the burning velocity of iso-octane and n-heptane with the counterflow flame technique (ambient pressure and different temperatures and equivalence ratios), Chong et al. (2011, [19]) obtained the burning velocity using the jet-wall stagnation flame configuration and the particle imaging velocimetry and Dirremberger et al. [20] used a perforated plate burner at 1 atm and 358 K, for iso-octane, h-heptane and the toluene mixture.

In this work, the combustion process of iso-octane is studied and characterized in a constant volume combustion bomb with spherical geometry. After that, the burning velocities of n-heptane and a mixture of n-heptane and toluene are experimentally obtained at engine like conditions. This experimental facility has been used and validated before for the determination of the burning velocity of gaseous fuels and is validated in this work for liquid fuels. The values obtained for the burning velocity are expressed as power laws of the pressure, temperature and equivalence ratio. The experimentally registered pressure is the input for a two-zone thermodynamic model used to determine the burning velocity of different fuel mixtures.

This paper presents results of the burning velocities for gasoline surrogates at a high pressure and temperature, i.e., engine like conditions. In the literature there are not many results for gasoline surrogate fuels, and present results expand the burning velocity database to different conditions and mixtures including toluene for a surrogate fuel as a binary mixture with n-heptane. In addition, the burning velocity values are important for the studies developing explosion protection. Additionally, surrogate fuels are currently used as fuels in simulations for advanced combustion engines (for example HCCI, homogeneous charge compression ignition engines, or PCCI, premixed charge compression ignition engines), which means it is necessary to obtain results for the burning velocity of these fuels.

2. Experimental Facility and Combustion Model

The experimental installation used in this work consisted of a test facility designed for the characterization and investigation of the combustion process of gaseous and liquid fuels. The main part of the facility is a spherical constant volume combustion bomb (CVCB) with an acquisition system to register the parameters during the combustion and obtain information about the flame development, and supply lines for the introduction of fuels in the combustion bomb. Fuel evaporation is required for liquid fuels in order to obtain a homogeneous mixture at the beginning of the combustion process. In Figure 1, a scheme of the experimental facility can be seen, where the CVCB is a spherical space made of stainless-steel 200 mm in diameter, pressure and temperature transducers and two optical accesses which are radial and horizontal to detect and study the chemiluminescence emitted by the flame. The CVCB has been designed to withstand pressures up to 40 MPa and temperatures up to 1073 K during the course of the combustion process. There are two electrodes inside the CVCB between which the spark is discharged to start the combustion at the geometrical center of the sphere, for more details see [21].

The initial conditions of pressure, temperature and fuel/air ratio were set up at the beginning of each combustion test. Liquid fuels were directly introduced in the combustion chamber and the air was added later with the corresponding line supply. A mixture time is necessary to get a homogeneous mixture inside the CVCB. Once the combustion was initiated, a spherical flame front propagates inside the combustion bomb compressing adiabatically and burning the fresh mixture. During the combustion process development, a piezoelectric transducer, Kistler 7063 type (maximum calibration error of 0.06%), registered the evolution of the pressure. This transducer was connected to a KISTLER 5018A1000 charge amplifier (maximum calibration error of 0.3%). The output signal of the charge amplifier was recorded on a Yokogawa DL750 Scopecorder (16 bits AD converter). The estimated error of the pressure acquisition is 0.36% over the measuring range. With the acquisition system it is possible to obtain values of up to 30,000 Hz. Additional details of the experimental facility and of the use of OH and CH chemiluminescence for flame characterization can be seen in Tinaut et al. [22].

The burning velocity was determined by the two-zone combustion analysis model, in which the main input was the temporal evolution of the pressure registered during the combustion, in addition to the initial values of the fuel-composition and mass of the fuel blend.

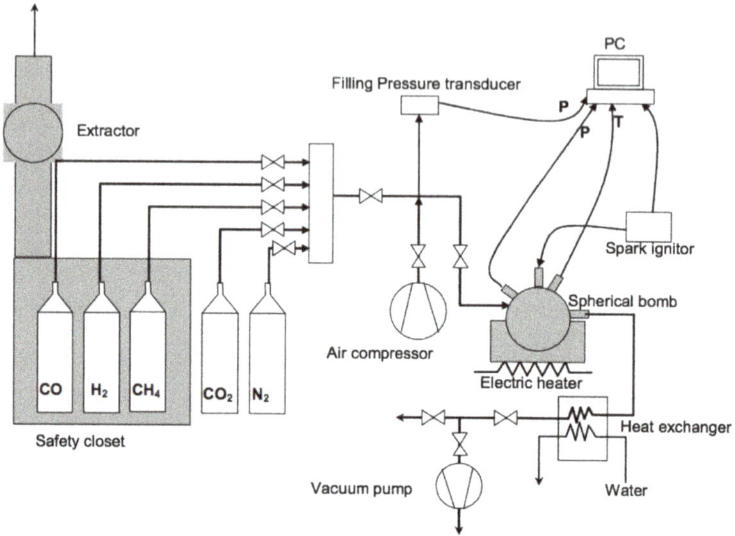

Figure 1. Experimental facility.

The thermodynamic combustion model considers the division of the combustion chamber in two different zones: the burned zone (designated with a b subscript) and unburned zone (designated with a ub subscript). In each zone, there were applied conservation and ideal gas equations [23–25]. Once the combustion has been initiated by the spark plug, a flame front appears and the spherical flame moves concentrically with the vessel walls, as the unburned or fresh mixture transforms into burned mixture. The outputs of this model are the temperatures of both zones, the burned mass fraction, the flame front surface and the burning velocity, among other variables [23]. The burning velocity, Cc, was determined from the mass burning rate ($\dot{m}_b$), the flame front surface (A_f) and from the unburned density (ρ_{ub}) according to the next expression:

$$C_c = \frac{\dot{m}_b}{\rho_{ub} A_f}, \tag{1}$$

As indicated in [26], some parts of the burning velocity plots were discarded to calculate the laminar burning velocity: some part of the initial points (due to the effects of the spark plug on the combustion onset, numerical oscillations and stretch rate) and some part of the final points (the autoignition processes and cellularity effects due to instabilities). A cellular combustion is obtained under certain conditions in which a cellular flame front ends in an apparent burning velocity higher than the laminar one when both velocities are referred to a smooth flame front—they can also be referred to the spherical one. The last part of the combustion process also had to be discarded because the flame front reached the CVCB wall and it could disturb the free flame development. The unburned temperature could also be locally perturbed due to heat transmission, and the hypothesis of adiabatic compression for the unburned gas temperature calculation will stop being fulfilled and some buoyancy effects may appear.

3. Results

Initially, the methodology for the determination of the burning velocity was presented and validated for liquid fuels. Iso-octane is the fuel used because of the existence of the extensive literature studying combustion. After that, the burning velocity of n-heptane and a mixture of 50% n-heptane/50%

toluene are obtained for different conditions of the pressure, temperature (engine like conditions) and equivalence ratio, and are compared with literature data.

3.1. Validation of the Methodology for Liquid Fuels: Iso-Octane

In this section, the methodology for the determination of the burning velocity of liquid fuels is validated using iso-octane as a fuel. This methodology has been previously validated with gaseous fuels.

The influence of pressure and temperature is studied in a stoichiometric mixture of iso-octane and air for the conditions shown in Figure 2, with an initial pressure between 0.1 and 0.5 MPa and initial temperature between 360 and 450 K. In Figure 3, the pressure evolution versus the flame front radius in a combustion process of stoichiometric iso-octane is plotted. In Figure 3, it is possible to see that the pressure reaches 50% of its maximum value when the flame front is close to the wall of the CVCB (94.2 mm radius). In Figure 4, the temporal evolution of the pressure (Figure 4a, experimental values) and burning velocity (Figure 4b, obtained by means of the 2-Z model) is plotted versus unburned temperature, for stoichiometric combustions of iso-octane at different initial conditions. It can be seen in both Figures that the burning velocity reduces as the initial pressure is increased.

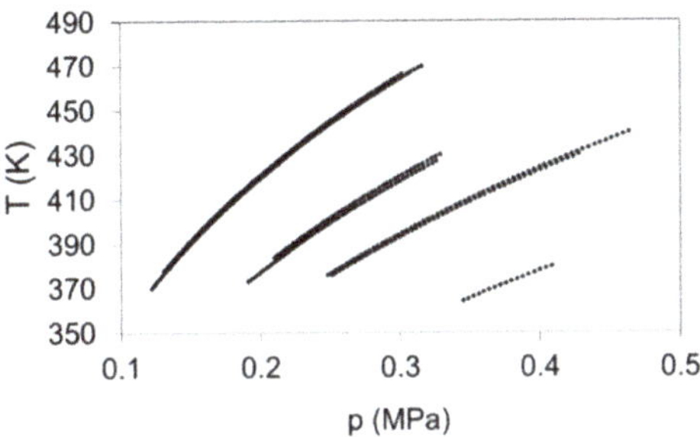

Figure 2. Pressure–temperature lines of adiabatic compression of iso-octane used for the calculation of the burning velocity for stoichiometric conditions.

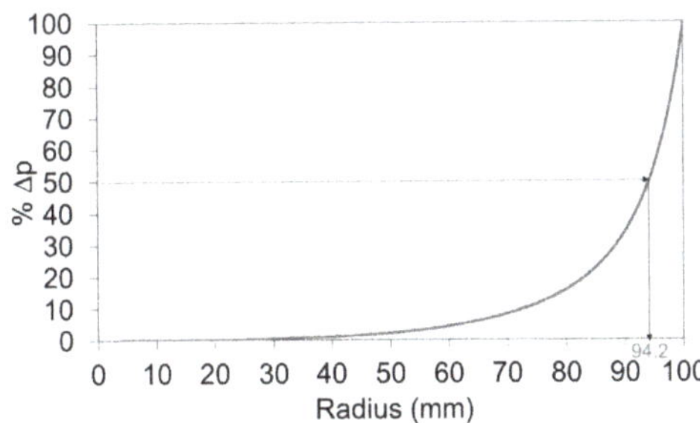

Figure 3. Pressure increase versus the flame front radius in a combustion of iso-octane, with a stoichiometric fuel/air equivalence ratio, 0.1 MPa and 353 K initial conditions.

The range of valid data used for the burning velocity determination excludes initial data affected by the ignition process (a value of the mass fraction burned, equal to 0.05, is used as the initial validity point for the burning velocity). It also excludes the final points and data affected by instabilities and the autoignition process. Burning velocities are obtained with the two-zones thermodynamic model. The burning velocity shows a smooth increment, except for the lower pressure combustion where it is possible to appreciate a bump in the curve, caused by the instabilities of the flame front. Additionally, autoignition phenomena may arise, generating oscillations in the pressure line and a characteristic sharp bump in the burning velocity plots (as can be seen in some experiments presented in following sections).

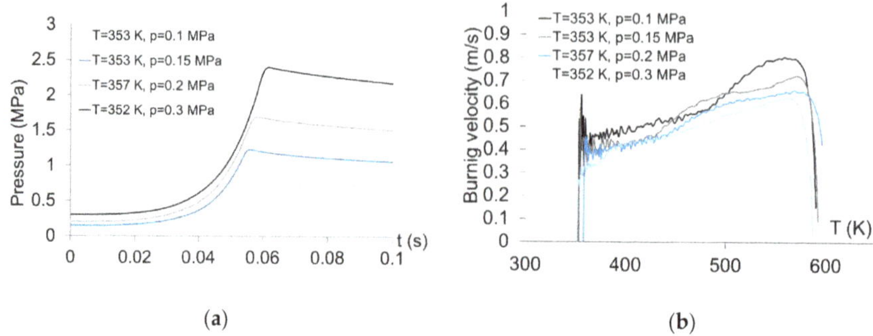

Figure 4. (**a**) Pressure; (**b**) burning velocity. The temporal evolution of the pressure and burning velocity of iso-octane versus the unburned temperature for stoichiometric mixtures with varying initial pressures.

The burning velocity values shown in Figure 4b can be fitted to a power law correlation as a function of pressure and temperature, of the type proposed by Metghalchi–Keck [8], as seen in Equation (2), where C_c is the burning velocity in m/s, C_{co} is the reference burning velocity at the T_o and p_o, α and β are the exponents of temperature and pressure to be determined. The value of the correlation coefficient R^2 is 0.96 and the standard error of estimation is 4%.

$$C_c = C_{co}\left(\frac{T}{T_o}\right)^{\alpha}\left(\frac{p}{p_o}\right)^{\beta}, \tag{2}$$

$$C_c(iso-octane) = 0.32\left(\frac{T}{300}\right)^{1.91}\left(\frac{p}{0.1}\right)^{-0.25}, \tag{3}$$

In Table 1 the correlation obtained in the present work is presented together with the expressions obtained by Gülder [9], Bradley et al. [7], Metghalchi et al. [10], Galmiche et al. [15] and Müller et al. [11] for the stoichiometric fuel/air equivalence ratio. In Müller's expression, T_o is the temperature of the interior zone, which marks the transition between the reaction zone and the unburned zone (the temperature that limits chain brain and chain break reactions), Y_f is the unburned fuel mass fraction and Tad is the adiabatic temperature.

Table 1. Correlations for the burning velocity of stoichiometric iso-octane.

Gülder [6]	$C_c = 0.45\left(\frac{T}{300}\right)^{1.56}\left(\frac{p}{0.1}\right)^{-0.22}$
Metghalchi_Keck [8]	$C_c = 0.25\left(\frac{T}{300}\right)^{2.18}\left(\frac{p}{0.1}\right)^{-0.16}$
Present work	$C_c = 0.32\left(\frac{T}{300}\right)^{1.91}\left(\frac{p}{0.1}\right)^{-0.25}$
Bradley et al. [7]	$C_c = 0.40\left(\frac{T}{300}\right)^{1.01}\left(\frac{p}{0.1}\right)^{-0.28}$
Galmiche et al. [9]	$C_c = 0.56\left(\frac{T}{423}\right)^{1.89}\left(\frac{p}{0.1}\right)^{-0.26}$
Müller et al. [10]	$C_c = 2.93\times10^3\left(\frac{p}{3.80\times10^7}\right)^{(1.22\times10^{-3})} Y_f^{0.56}\left(\frac{T_{ad}-T^o}{T_{ad}-T}\right)^{2.52}\left(\frac{T}{T_o}\right)$

In Figure 5a, the burning velocity of the stoichiometric mixtures of iso-octane and air is plotted versus the unburned temperature for an initial pressure of 0.5 MPa. The line obtained in this work is placed at the center of the plot, with a slope similar to the one obtained by Müller et al., in spite of the fact that the Müller expression is different and they obtain it from a program with an extensive reaction mechanism. Galmiche et al. [9] measured the flame velocities of spherically expanding flames using a combustion chamber equipped with a high-speed shadowgraph system. The differences between the expression of Bradley et al. [7] and Metghalchi et al. [8] are due to the fact that they considered all the data obtained during the combustion processes without any distinction between the laminar and cellular zone, which explains the higher slope of the Metghalchi correlation. Gülder [6] obtained the expression for the burning velocity in a combustion bomb using ionization sensors in the combustion zone, leading to higher values.

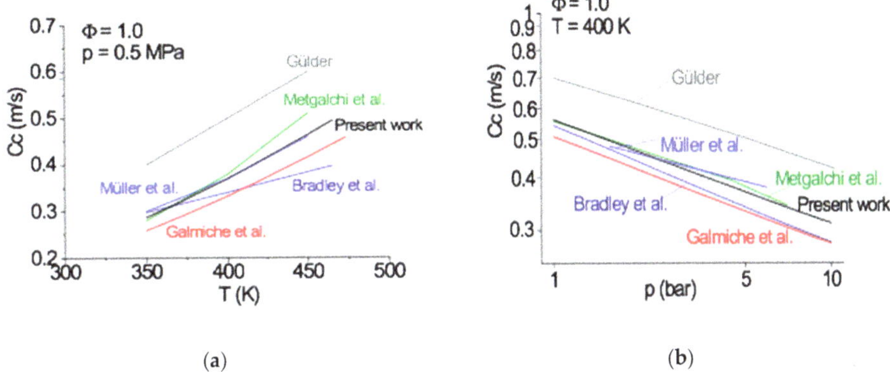

(a) (b)

Figure 5. The burning velocity for stoichiometric iso-octane versus temperature (**a**) and pressure (**b**).

In Figure 4b, the burning velocity versus the pressure is represented for an initial temperature of 400 K. The tendencies of the present work results are similar to those explained before: the slope is similar to the one obtained by Müller et al. [10]. The differences with the correlations obtained by Gülder [6] and Bradley et al. [7] are due to a different measurement method being used, while the differences with the Metghalchi and Keck [8] values are due to the range of considered data.

The stoichiometric iso-octane burning velocity is represented in Figure 6 versus the temperature and pressure, for the expressions shown in Table 1. The highest values of the burning velocity are obtained for elevated temperatures and low pressures, in all the cases. In Figure 6, all the correlations are represented together in order to compare them more clearly. In addition, some experiments with varying equivalence ratios were developed to compare with the literature data, as seen in Figure 7. The values obtained in the present work (black points) are compared with the values of Galmiche et al. (2012, [9]), Metghalchi and Keck (1982, [8]), Varea et al. (2013, [11]), Marshall et al. (2011, [12]) and Gülder (1982, [6]). It can be seen that the values obtained in the present work agree with the rest of the data.

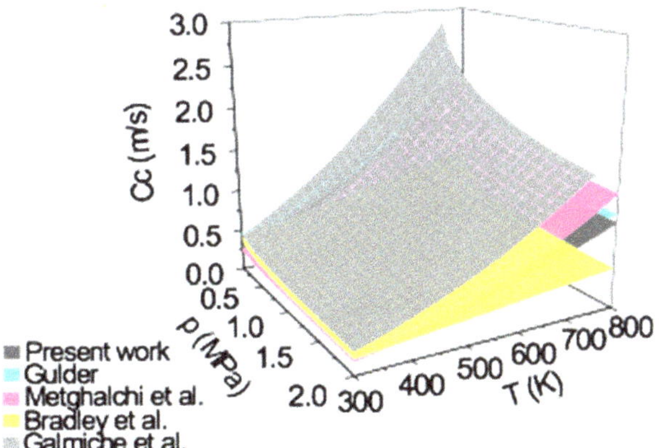

Figure 6. The burning velocity of iso-octane as a function of temperature and pressure for the stoichiometric equivalence ratio. Comparison between the expressions obtained by Gülder [6], Metghalchi and Keck [8], Bradley et al. [7], Galmiche et al. [9] and those obtained in the present work.

As a conclusion of this section, it is possible to say that the methodology for the determination of the burning velocity of liquid fuels in the CVCB has been validated and the results obtained for other fuels will be presented in the following sections.

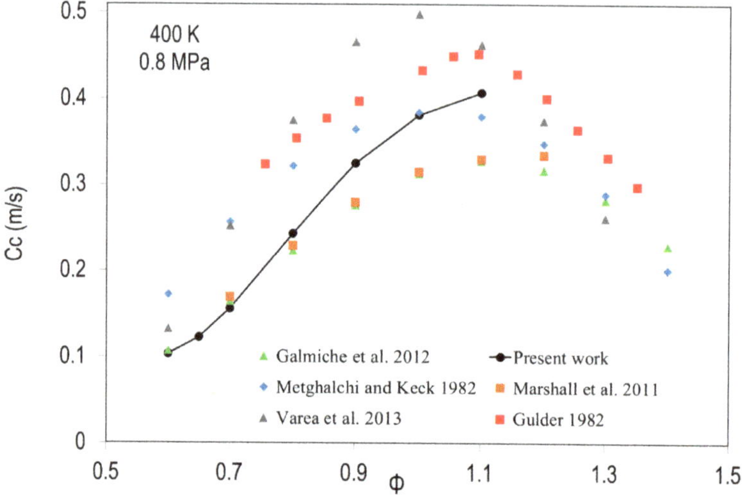

Figure 7. Comparison of the burning velocity of iso-octane versus the equivalence ratio, for 0.8 MPa and 400 K of the initial conditions.

3.2. *Burning Velocity of n-Heptane*

In this section, the results for the burning velocity of n-heptane/air mixtures in the CVCB are presented for the different initial conditions of the pressure and temperature. The initial conditions are chosen to be able to reach the range of pressure and temperature obtained by the internal combustion in engines during the progress of the combustion, as can be seen in Figure 8a. The objective of these experiments is to characterize the burning velocity of n-heptane through correlations as a function of pressure and temperature and the fuel/air equivalence ratio, Equation (2).

3.2.1. Results of n-Heptane for a Stoichiometric Mixture

A pressure–temperature diagram of all the data used to determine the burning velocity of n-heptane in stoichiometric conditions is shown in Figure 8a. In this figure, points represent the adiabatic evolution of the pressure and temperature during the combustion process inside the CVCB, starting from the initial conditions of pressure and temperature. Figure 8a delimits the validity range of the obtained correlation, with an initial pressure between 0.1 and 1.2 MPa and initial temperature between 370 and 650 K.

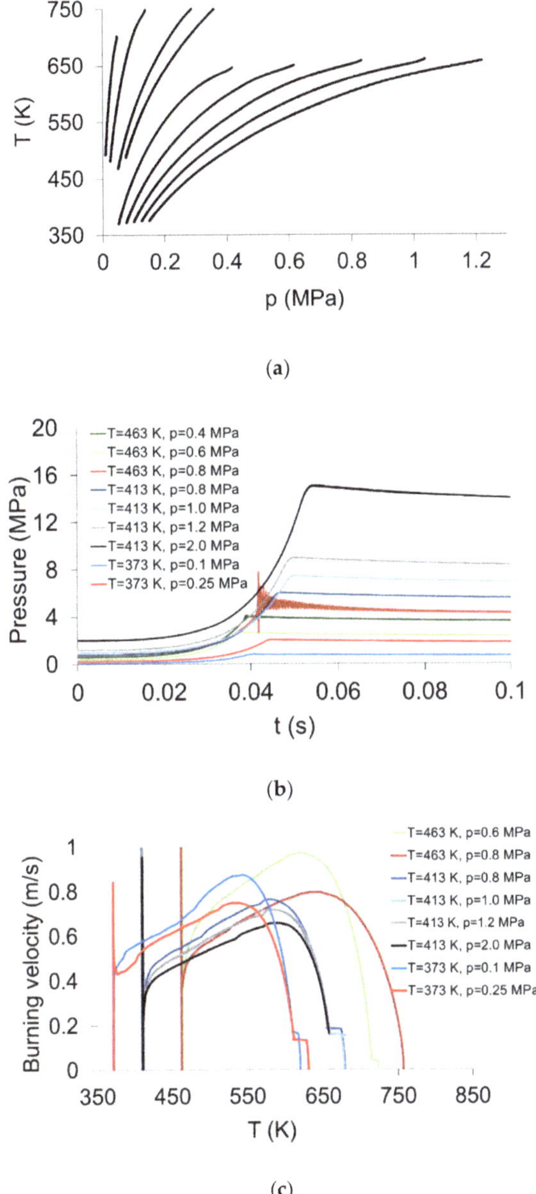

Figure 8. (**a**) Pressure–temperature diagram; (**b**) pressure; (**c**) burning velocity. The temporal evolution of the pressure and burning velocity of n-heptane versus the unburned temperature for stoichiometric combustions.

Once the mixture is ignited at the center of the combustion chamber, the temporal evolution of the pressure during the combustions for stoichiometric mixtures of n-heptane/air for different initial conditions are plotted in Figure 8b (combustions are ordered by the initial temperature). Pressure starts growing until the flame front touches the CVCB wall, and the pressure curves reach their maximum value. As the initial pressure increases, for the same initial temperature, combustions are slower and the rise of the curve is smoother, and the final pressure increases too, due to the adiabatic compression inside the CVCB. For the combustion of 463 K and 0.8 MPa (turquoise line), the autoignition of the n-heptane mixture is reached, causing vibrations in the pressure curve. Additionally, a low autoignition process is observed in the pressure curve for 463 K and 0.4 MPa (dark blue line). Autoignition takes place inside the combustion bomb when the autoignition conditions are reached during the adiabatic compression, and a part of the unburned mixture autoignites. Burning velocities (obtained with the thermodynamic model) versus unburned are represented in Figure 8c, where it is possible to see that as the pressure increases, for a given temperature, the burning velocity decreases. In contrast, when the pressure is constant and the temperature increases, the burning velocity increases with temperature. Results obtained for n-heptane are qualitatively similar to the case of iso-octane.

A Metghalchi–Keck type correlation for the stoichiometric combustion of n-heptane is shown in Equation (4), where the value of the correlation coefficient R^2 is 0.989 and the standard error of estimation is 2%.

$$Fr = 1 \rightarrow Cc\left(\frac{m}{s}\right) = 0.35\left(\frac{T}{T_0}\right)^{2.21}\left(\frac{p}{p_o}\right)^{-0.21}, \quad (4)$$

3.2.2. Results of n-Heptane for a 0.9 Fuel/Air Equivalence Ratio

The validity ranges of the data used for the determination of the burning velocity correlation of n-heptane with a 0.9 equivalence ratio are shown in thick trace on the temperature–pressure evolution lines of Figure 9a, with an initial pressure between 0.1 and 1.2 MPa and an initial temperature between 360 and 650 K. The temporal evolution of pressure in the combustions of n-heptane are shown in Figure 9b, and the corresponding burning velocities are plotted in Figure 9c.

The general trend is that the burning velocity increases with the initial temperature, and decreases with the initial pressure. It can be seen that in some of the experimental cases, characterized by high initial temperatures and/or pressures (450 K–1.7 MPa—yellow line, 451 K–2.0 MPa—blue line, and 474 K–0.8 MPa—red line), appear some oscillations in the pressure curve due to autoignition process. The corresponding burning velocity plots show a significant increment in the apparent burning velocity.

The explanation of this behavior is the onset of autoignition processes, which lead to a more violent and faster combustion process than the previous laminar premixed combustion. This autoignition behavior did not appear in the case of iso-octane (in the tested experimental range), which agrees with the well-known difference between the octane number of iso-octane and n-heptane. When the autoignition takes place in the fresh mixture, the diagnostic model gives very high values of the apparent burning velocity, because during this process some autoignition reactions play an important role in the unburned zone with a flame front, which are not geometrically identified. This behavior does not appear when the equivalence ratio is stoichiometric.

The obtained correlation with the data represented in Figure 9c is the following, where the value of the correlation coefficient R^2 is 0.973 and the standard error of estimation is 7%.

$$Fr = 0.9 \rightarrow Cc\left(\frac{m}{s}\right) = 0.33\left(\frac{T}{T_0}\right)^{2.73}\left(\frac{p}{p_o}\right)^{-0.30}, \quad (5)$$

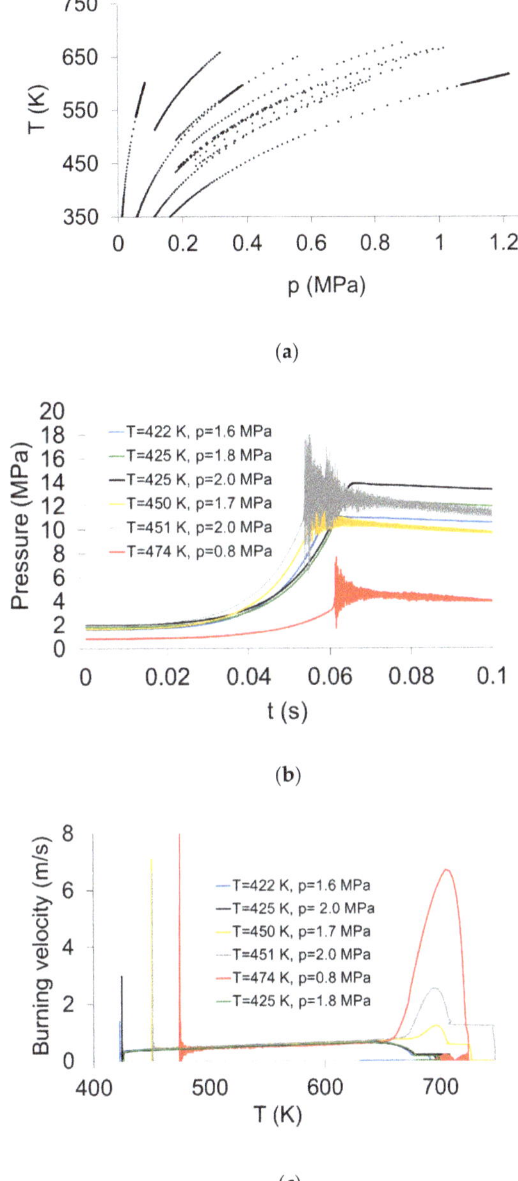

Figure 9. (**a**) Pressure–temperature diagram; (**b**) pressure; (**c**) burning velocity. Temperature–pressure map, the temporal evolution of pressure and burning velocity versus the unburned temperature for the combustions of n-heptane with a 0.9 of equivalence ratio.

3.2.3. Results of n-Heptane for a 0.8 Fuel/Air Equivalence Ratio

In the case of an equivalence ratio of 0.8, the observed experimental behavior of the combustion process of n-heptane is qualitatively very similar to the 0.9 case. Autoignition appears in some combinations of initial high temperature and pressure, leading to very high apparent burning velocities in the latter part of the combustion process.

For brevity, only the final result is presented in the form of a burning velocity adjusted to a Metghalchi and Keck type correlation, Equation (6), where the value of the correlation coefficient R^2 is 0.984 and the standard error of estimation is 3%.

$$Fr = 0.8 \rightarrow Cc\left(\frac{m}{s}\right) = 0.27\left(\frac{T}{T_0}\right)^{2.77}\left(\frac{p}{p_o}\right)^{-0.33}, \quad (6)$$

To check the accuracy of the correlations proposed for n-heptane combustion (Equations (4)–(6)), they are compared with other authors' results. The values are compared with those obtained by Davis and Law (1998, [13]), Huang et al. (2004, [14]), Van Lipzig et al. (2011, [15]), Sileghem et al. (2013, [16]), Marshall et al. (2011, [12]), Kwon et al. (2000, [17]), Kumar et al. (2007, [18]), Chong et al. (2011, [19]) and Dirremberger et al. [20] in Figure 10, for ambient conditions of pressure and temperature. It is possible to see a good agreement between results of this work and the other authors' results.

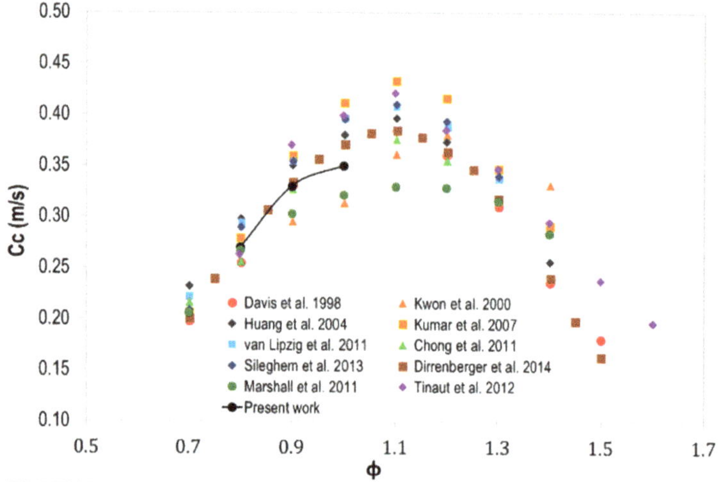

Figure 10. Comparison between the n-heptane burning velocities obtained in this work (black points) with the ones obtained by references [13].

3.3. Burning Velocity of a 50% n-Heptane and 50% Toluene Blend

In this section, results for the burning velocity of a blend of 50% n-heptane/50% toluene, in mass, is obtained in the CVCB for different initial conditions of pressure, temperature and equivalence ratios of 0.8 and 1.0 (right). This mixture is considered a surrogate fuel, where the toluene represents the aromatic content. The validity range for the stoichiometric blend is an initial pressure between 0.1 and 0.9 MPa and an initial temperature between 450 and 700 K, and for the 0.8 equivalence ratio, an initial pressure ranging from 0.1 to 1.0 MPa and an initial temperature from 360 to 700 K.

Figure 11 shows the pressure–temperature map, temporal evolution of pressure and burning velocity versus the unburned temperature, for combustions of 50% n-heptane/50% toluene for stoichiometric (left) and 0.8 (right) equivalence ratios. As in the case of pure n-heptane, for some initial conditions of pressure and temperature some oscillations can be observed in the pressure plots, associated to the onset of autoignition. In parallel, the apparent burning velocity increases suddenly and strongly after autoignition appears. There are more cases of autoignition apparition for the lean (0.8) mixture than for the stoichiometric mixture. This is a result of the trade-off between two competing processes: laminar combustion, which needs a time equivalent to the chamber radius over the burning velocity, and the autoignition process, with a delay time that depends on the pressure and temperature history of the air–fuel mixture experiments. Since the mixture is leaner when the laminar velocity is smaller, the time needed to complete the flame length is longer, thus making it bigger than the autoignition delay time and causing a generalized combustion of the remaining unburned mixture. This so-called abnormal combustion leads to a sharp increase in pressure, sometimes associated to secondary pressure waves that explain the oscillations in the pressure plots. When the two-zone thermodynamic model finds these sharp pressure increments, the apparent burning velocity computed as a result is very high, although not necessarily with a physical fundament.

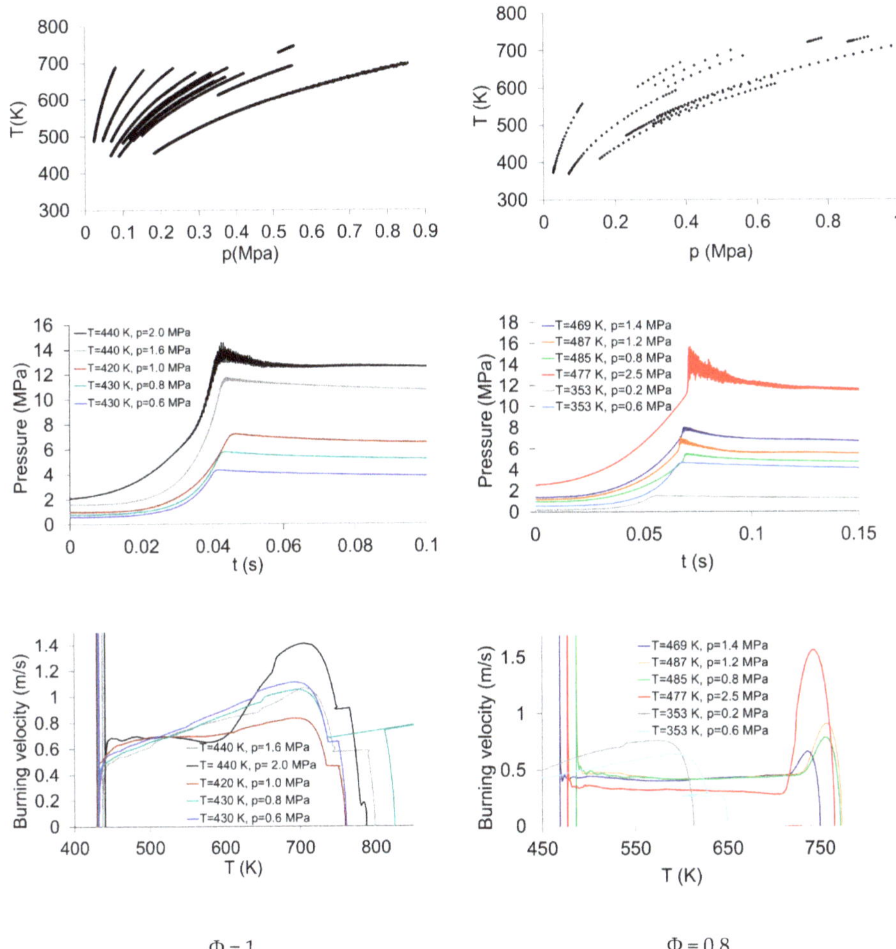

Figure 11. A temperature–pressure map, the temporal evolution of pressure and burning velocity versus temperature, for the mixture of 50% n-heptane/50% toluene, for different initial conditions and equivalence ratios (1.0-left, 0.8-right).

The burning velocities obtained with the two-zone thermodynamic model are adjusted to a correlation with the pressure and temperature shown in Equations (7) and (8) (for stoichiometric and 0.8 equivalence ratios, respectively). The value of the correlation coefficient R^2 is 0.950 and 0.976 and the standard error of estimation is 5% and 6%, for the stoichiometric and 0.8 equivalence ratios.

$$Fr = 1 \rightarrow Cc\left(\frac{m}{s}\right) = 0.31\left(\frac{T}{T_0}\right)^{1.95}\left(\frac{p}{p_o}\right)^{-0.103}, \quad (7)$$

$$Fr = 0.8 \rightarrow Cc\left(\frac{m}{s}\right) = 0.33\left(\frac{T}{T_0}\right)^{2.19}\left(\frac{p}{p_o}\right)^{-0.24} \quad (8)$$

4. Conclusions

A spherical constant volume combustion bomb has been used for the determination of the burning velocity of liquid fuels: iso-octane, n-heptane and a mixture with toluene. A two-zone thermodynamic combustion model allows for the analysis of the pressure register during the combustion and to obtain some significant properties during the combustion process, allowing us to characterize the physical and thermochemical properties of the combustion, as can be done with the burning velocity, mass burned fraction, temperatures, etc.

The burning velocity of three liquid fuels: iso-octane, n-heptane and a blend of 50% n-heptane/50% toluene in mass, has been obtained for the elevated conditions of pressure and temperature (engine like conditions) and diverse equivalence ratios. The burning velocities of n-heptane and iso-octane have been compared with data obtained in the literature, for the same conditions, obtaining a good agreement with them. The burning velocities are expressed as power law correlations of pressure and temperature for a given fuel/air equivalence ratio. For the three fuels studied, the burning velocity enhances with temperature (with exponents of 1.91 in the case of iso-octane and 2.21 in the case of n-heptane) during the combustion process. The dependence with pressure is the contrary—the burning velocity decreases with the increment of pressure (with exponents between 0.25 for iso-octane and 0.33 for n-heptane, with values between 0.10 and 0.23 for the 50% n-heptane/50% toluene mixture).

The traces of pressure of iso-octane combustion are smooth, as well as the corresponding burning velocities. However, during the n-heptane combustion, some oscillations are detected in the pressure and burning velocity curves, especially for lean mixtures, which is due to the onset of autoignition processes. The blend of n-heptane and toluene shows qualitatively the same behavior of n-heptane, with strong increments in the apparent burning velocity curves once autoignition appears.

Author Contributions: Conceptualization, M.R. and F.V.T.; investigation, M.R. and A.C.; methodology, M.R.; supervision, F.V.T.; validation, M.R. and A.C.; writing—original draft, M.R.; writing—review and editing, M.R., F.V.T. All authors have read and agreed to the published version of the manuscript.

Funding: This research was funded by the Spanish government of science and innovation with the research project, PID2019-106957RB-C22.

Acknowledgments: This work was conducted by the Research group in Engines and Renewable Energies (MYER) from the University of Valladolid in Spain under the framework of the PID2019-106957RB-C22 research project.

Conflicts of Interest: The authors declare no conflict of interest.

References

1. Egolfopoulos, F.N.; Cho, P.; Law, C.K. Laminar flame speeds of methane-air mixtures under reduced and elevated pressures. *Combust. Flame* **1989**, *76*, 375–391. [CrossRef]
2. Konnov, A.A.; Mohammad, A.; Kishore, V.R.; Kim, N.I.; Prathap, C.; Kumar, S. A comprehensive review of measurements and data analysis of laminar burning velocities for various fuel+ air mixtures. *Prog. Energy Combust. Sci.* **2018**, *68*, 197–267. [CrossRef]
3. Bhavani Shankar, V.S.; Sajid, M.; Al-Qurashi, K.; Atef, N.; Alkhesho, I.; Ahmed, A.; Chung, S.; Roberts, W.; Morganti, K.; Sarathy, M. *Primary Reference Fuels (PRFs) as Surrogates for Low Sensitivity Gasoline Fuels*; SAE International: Warrendale, PA, USA, 2016.
4. Mueller, C.J.; Cannella, W.J.; Bruno, T.J.; Bunting, B.; Dettman, H.D.; Franz, J.A.; Huber, M.L.; Natarajan, M.; Pitz, W.J.; Ratcliff, M.A.; et al. Methodology for Formulating Diesel Surrogate Fuels with Accurate Compositional, Ignition-Quality, and Volatility Characteristics. *Energy Fuels* **2012**, *26*, 3284–3303. [CrossRef]

5. Ra, Y.; Reitz, R.D. A reduced chemical kinetic model for IC engine combustion simulations with primary reference fuels. *Combust. Flame* **2008**, *155*, 713–738. [CrossRef]
6. Gülder, Ö.L. Laminar burning velocities of methanol, ethanol and isooctane-air mixtures. *Symp. Int. Combust.* **1982**, *19*, 275–281. [CrossRef]
7. Bradley, D.; Hicks, R.; Lawes, M.; Sheppard, C.; Woolley, R. The measurement of laminar burning velocities and Markstein numbers for iso-octane–air and iso-octane–n-heptane–air mixtures at elevated temperatures and pressures in an explosion bomb. *Combust. Flame* **1998**, *115*, 126–144. [CrossRef]
8. Metghalchi, M.; Keck, J.C. Burning velocities of mixtures of air with methanol, isooctane, and indolene at high pressure and temperature. *Combust. Flame* **1982**, *48*, 191–210. [CrossRef]
9. Galmiche, B.; Halter, F.; Foucher, F. Effects of high pressure, high temperature and dilution on laminar burning velocities and Markstein lengths of iso-octane/air mixtures. *Combust. Flame* **2012**, *159*, 3286–3299. [CrossRef]
10. Müller, U.; Bollig, M.; Peters, N. Approximations for burning velocities and Markstein numbers for lean hydrocarbon and methanol flames. *Combust. Flame* **1997**, *108*, 349–356. [CrossRef]
11. Varea, E.; Modica, V.; Renou, B.; Boukhalfa, A.M. Pressure effects on laminar burning velocities and Markstein lengths for Isooctane–Ethanol–Air mixtures. *Proc. Combust. Inst.* **2013**, *34*, 735–744. [CrossRef]
12. Marshall, S.P.; Taylor, S.; Stone, C.R.; Davies, T.J.; Cracknell, R.F. Laminar burning velocity measurements of liquid fuels at elevated pressures and temperatures with combustion residuals. *Combust. Flame* **2011**, *158*, 1920–1932. [CrossRef]
13. Davis, S.; Law, C. Determination of and fuel structure effects on laminar flame speeds of C1 to C8 hydrocarbons. *Combust. Sci. Technol.* **1998**, *140*, 427–449. [CrossRef]
14. Huang, Y.; Sung, C.; Eng, J. Laminar flame speeds of primary reference fuels and reformer gas mixtures. *Combust. Flame* **2004**, *139*, 239–251. [CrossRef]
15. Van Lipzig, J.; Nilsson, E.; De Goey, L.; Konnov, A.A. Laminar burning velocities of n-heptane, iso-octane, ethanol and their binary and tertiary mixtures. *Fuel* **2011**, *90*, 2773–2781. [CrossRef]
16. Sileghem, L.; Alekseev, V.A.; Vancoillie, J.; Van Geem, K.M.; Nilsson, E.J.K.; Verhelst, S.; Konnov, A.A. Laminar burning velocity of gasoline and the gasoline surrogate components iso-octane, n-heptane and toluene. *Fuel* **2013**, *112*, 355–365. [CrossRef]
17. Kwon, O.; Hassan, M.; Faeth, G. Flame/stretch interactions of premixed fuel-vapor/O/N flames. *J. Propuls. Power* **2000**, *16*, 513–522. [CrossRef]
18. Kumar, K.; Freeh, J.; Sung, C.; Huang, Y. Laminar flame speeds of preheated iso-octane/O2/N2 and n-heptane/O2/N2 mixtures. *J. Propuls. Power* **2007**, *23*, 428–436. [CrossRef]
19. Chong, C.T.; Hochgreb, S. Measurements of laminar flame speeds of liquid fuels: Jet-A1, diesel, palm methyl esters and blends using particle imaging velocimetry (PIV). *Proc. Combust. Inst.* **2011**, *33*, 979–986. [CrossRef]
20. Dirrenberger, P.; Glaude, P.-A.; Bounaceur, R.; Le Gall, H.; Da Cruz, A.P.; Konnov, A.; Battin-Leclerc, F. Laminar burning velocity of gasolines with addition of ethanol. *Fuel* **2014**, *115*, 162–169. [CrossRef]
21. Tinaut, F.V.; Reyes, M.; Giménez, B.; Pastor, J.V. Measurements of OH* and CH* Chemiluminescence in Premixed Flames in a Constant Volume Combustion Bomb under Autoignition Conditions. *Energy Fuels* **2011**, *25*, 119–129. [CrossRef]
22. Tinaut, F.V.; Melgar, A.; Giménez, B.; Reyes, M. Characterization of the combustion of biomass producer gas in a constant volume combustion bomb. *Fuel* **2010**, *89*, 724–731. [CrossRef]
23. Tinaut Fluixá, F.; Giménez Olavarría, B.; Iglesias Hoyos, D.; Lawes, M. Experimental Determination of the Burning Velocity of Mixtures of n-Heptane and Toluene in Engine-like Conditions. *FlowTurbul. Combust.* **2012**, *89*, 183–213. [CrossRef]
24. Horrillo, A. *Utilization of Multi-Zone Models for the Prediction of the Pollutant Emissions in the Exhaust Process in Spark Ignition Engines*; University of Valladolid: Valladolid, Spain, 1998.
25. Heywood, J.B. *Internal Combustion Engine Fundamentals*; Hill, M., Ed.; McGraw-Hill: New York, NY, USA, 1988.
26. Reyes, M.; Tinaut, F.V.; Horrillo, A.; Lafuente, A. Experimental characterization of burning velocities of premixed methane-air and hydrogen-air mixtures in a constant volume combustion bomb at moderate pressure and temperature. *Appl. Therm. Eng.* **2018**, *130*, 684–697. [CrossRef]

Article

The Potential of Wobble Plate Opposed Piston Axial Engines for Increased Efficiency

Paweł Mazuro and Barbara Makarewicz *

Department of Aircraft Engines, Faculty of Power and Aeronautical Engineering, Warsaw University of Technology, 00-665 Warsaw, Poland; pawel.mazuro@pw.edu.pl
* Correspondence: barbara.makarewicz@pimet.com.pl

Received: 29 September 2020; Accepted: 22 October 2020; Published: 26 October 2020

Abstract: Recent announcements regarding the phase out of internal combustion engines indicate the need to make major changes in the automotive industry. Bearing in mind this innovation trend, the article proposes a new approach to the engine design. The aim of this paper is to shed a new light on the forgotten concept of axial engines with wobble plate mechanism. One of their most important advantages is the ease of use of the opposed piston layout, which has recently received much attention. Based on several years of research, the features determining the increase in mechanical efficiency, lower heat losses and the best scavenging efficiency were indicated. Thanks to the applied Variable Compression Ratio (VCR), Variable Angle Shift (VAS) and Variable Port Area (VPA) systems, the engine can operate on various fuels in each of the Spark Ignition (SI), Compression Ignition (CI) and Homogeneous Charge Compression Ignition (HCCI)/Controlled Auto Ignition (CAI) modes. In order to quantify the potential of the proposed design, an initial research of the newest PAMAR 4 engine was presented to calculate the torque curve at low rotational speeds. The achieved torque of 500 Nm at 500 rpm is 65% greater than the maximum torque of the OM 651 engine of the same 1.8 L capacity. The findings lead to the conclusion that axial engines are wrongfully overlooked and can significantly improve research on new trends in pollutant elimination.

Keywords: axial engines; wobble plate; opposed piston engine; uniflow scavenging; variable compression ratio; variable valve timing; downsizing; downspeeding; multifuel potential

---

1. Introduction

In the light of climate change and further negative ecological forecasts, the lawmakers from all over the world are trying to reduce the impact of human activity on the environment. In case of the engine industry, they aim at limiting the emission of harmful gases into the atmosphere. The regulations planned to be introduced in some countries are very strict. A recent review of this matter [1] reports, that the most ambitious country (Norway) projects that all cars and light vans sold from 2025 onward are going to be zero-emission vehicles. Denmark set a target to stop the sales of new gasoline and diesel cars by 2030. That same year, Iceland is going to outlaw the registration of such vehicles and Ireland proposed to ban the sale of new fossil-fuel cars by this date. According to the International Council of Clean Transportation (ICCT) briefing, ten European countries have made commitments regarding the phase-out of combustion engine passenger cars by 2040. Related to this are announcements from major manufacturers regarding the planned market launch of a significant amount of electric cars. Investments of tens of billions of euros are also planned with regards to the electrification of their vehicle portfolio.

Validity of departing from internal combustion engines is a subject of separate publications and will not be discussed in this article. As others have highlighted [2], electricity can be produced in various ways that may be more or less environmentally friendly. The issues that must be resolved are also the extraction of elements for the production of batteries and their storage at the end of their service life. Nevertheless, the abovementioned declarations of the engine manufacturers clearly show the willingness to introduce significant changes in their production profile. Bearing in mind the global trend for huge investments and innovation, the authors of this article would like to present a certain direction in the development of internal combustion engines, which may advance research aimed at reducing the air pollution. The paper takes a new look at the crank-piston mechanism and proposes a different design which could be a vital factor in emission control.

The subject of nearly two decades of research at the Warsaw University of Technology (WUT) is an axial engine (referred also as barrel engine) with a wobble-plate mechanism, working in an opposed piston layout. It is not a new concept, as the first engines of this type appeared in the late 19th and early 20th century. However, after the Second World War, axial engines haven't been getting much attention as they were pushed out of the market by technologically simpler engines with a crank mechanism [3]. For a long time, designers were unnecessarily averse to axial engines. It was caused by the opinion about gross errors in kinematics, which prevented their correct work [4]. The manufacturing issues were also a significant limitation.

In our opinion, this reasoning relies too heavily on the design capabilities of the last century. The technology has advanced considerably since the first axial engines were built-an example could be the spherical plain bearings, which are essential for the proper operation of many wobble plate mechanisms. The team at the Warsaw University of Technology has reassessed the benefits and problems that may arise during the operation of barrel engines. As the research shows, many hypotheses regarding the axial engines have been oversimplistic. This is probably the reason for the fact that the idea to return to this forgotten concept arose in a few research centers. Already at the end of the 20th century, there were attempts to classify and gather knowledge about axial engines with wobble plate mechanisms.

Publication [5] reviews US patents for wobble plate engines and their assessment based on several parameters related to their kinematics. The authors stated that as of the year of publication of their article (1986), there was no systematic knowledge about axial engines and methods of their design. In [4] attention was paid to the benefits of using wobble plate mechanisms, mainly the simplicity of using the Variable Compression Ratio system. The lack of experience in understanding the operation of such engines was again highlighted.

This knowledge gap was partially filled in the doctoral dissertation [3], where the kinematics of various mechanisms of the axial engines, with particular emphasis on the wobble plate, was analyzed in detail. The mechanisms with the lowest mechanical losses have been selected. Then, for over twenty years, during various projects, prototypes with the most promising kinematics were built. The aim of the article is to present the conclusions of many years of analysis and research on barrel engines in the context of their efficiency and environmental impact. The initial test results of the PAMAR 4 engine were presented, which clearly show that the barrel engines fit in the trends of modern automotive research with particular attention to downsizing and downspeeding.

Interest in axial engines is also arising in other research centers. The projects carried out include the Duke engine [6] from New Zealand and the Covaxe engine [7] from the UK.

1.1. Unconventional Internal Combustion Engine Designs

For years, engineers have seen the limitations of using classic internal combustion engines and they are trying to create alternative designs. One of the most interesting inventions are rotary engines, a comprehensive review of which was carried out in the publication [8]. This type of engine can be divided into three main types: vane, toothed gear and oscillatory engines, the most famous example of which is the Wankel design. The undoubted advantages of these engines include a compact design, barrel shape, and a significantly limited number of parts (no complicated timing system). One of the latest constructions of this type is the twin rotor piston engine, design of which is described in the publication [9]. The engine works on the principle of cooperation between quadrant vane pistons mounted on two oscillating rotors. It has been noted that this engine has a number of power strokes equal to the square of the number of vanes on each rotor. For the design presented in [9] with six pistons on one rotor, the engine will work with 36 power strokes per revolution. Rotary engines have unfortunately some fundamental disadvantages, such as the shape of the combustion chambers with a low volume to surface ratio which causes greater heat loss. The flame quenching in narrow spaces can cause increased emissions of hydrocarbons. There is also the so far unsolved problem of sealing between individual working chambers.

Another idea for reducing disadvantages of internal combustion engines is the free piston concept. This mechanism has no connecting rod or crankshaft-the movement of the piston is a result of gas forces from the combustion chamber and the reaction from a load device, such as an electrical power generator, gas compressor or oil pump. Such a system has many advantages-fewer parts and the possibility of integrating an electric generator directly in the engine with magnets in the piston and coil windings in the housing. Due to the completely different movement mechanism of the piston, no side piston force is generated. Simple implementation of a Variable Compression Ratio system is also possible. The patent review [10] showed that key automotive players are interested in this technology. General Motors, Toyota, Volvo, Ford, Honda and Mazda have filed patents for various free piston designs such as single piston, dual piston and opposed piston types. Research effort is motivated, among others, by the possibility of using the free piston engine as a power generator for hybrid electric vehicles. The future of this technology depends above all on solving one of the main challenges of the precise piston motion control. As stated in the review, there is not enough operational experience to predict the viability of free piston engines and develop satisfactory solutions for starting, continuous operation and engine cooling.

The abovementioned designs have many advantages, while the engineers will probably have to wait for further technology development to solve the challenges related to them. That is not the case with wobble plate engines, for which the current state of technology already allows solving the issues considered historically as defects. The historical outline of the construction of these engines is presented in the publications [11,12] and [3]. From the 1910s to the 1930s, designs that could be used in the military, aviation and automotive industries were developed around the world. Their classification is shown in Figure 1.

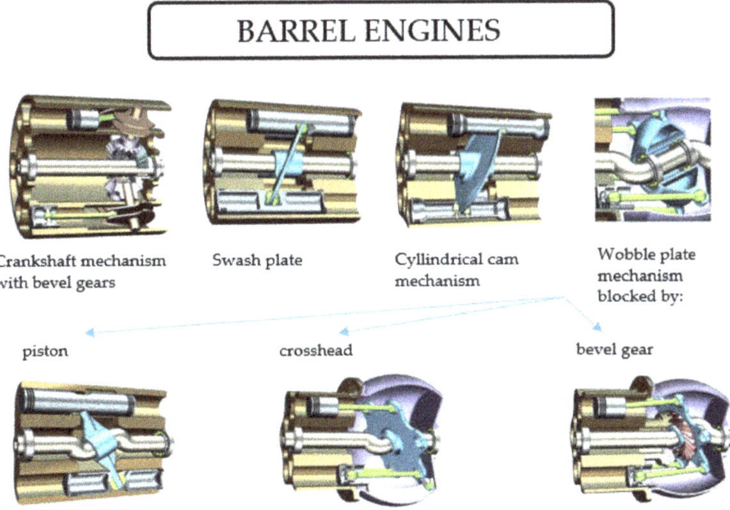

Figure 1. Classification of mechanisms used in axial engines [3].

The disadvantages mentioned in the descriptions of historical designs include: failure of lubrication, cooling, mechanical failures of wobble plate and pistons joints, high friction losses. The crankshaft mechanisms with bevel gears were abandoned due to an overly complicated mechanism and difficulties in servicing. Cylindrical cam engines are also a large group of mechanisms with the possibility of almost freely choosing the piston motion equation. According to the authors, however, such constructions should be abandoned in internal combustion engines, because they are characterized by greater side piston force, which has been shown later in this article.

As [11] states, some of the mechanisms have not been thoroughly tested because of changing the financing policy, other ready-made solutions appearing faster on the market or change of priorities during the war. That was the case with Almen wobble plate design (1917–1920), where the funding was cut after five prototypes were already designed. A highly developed state of design has reached also a German construction discovered after the Second World War by the Allies (1945). It is likely that the German designers were developing an idea previously presented in the Alfaro engine from the USA.

1.2. Challenges Concerning Introduction of Wobble Plate Engines to the Market

The article presents the concept of an unconventional engine, which, according to the authors, has the best chance of being an alternative to engines with a crankshaft. However, it cannot be ignored that the market launch of a new engine is quite an ambitious and difficult task. It presents engineers with many challenges that can temporarily be perceived as flaws and discourage both research teams and investors. The following subsection presents the issues that, according to the authors, constitute the greatest obstacles for further research on the discussed engines.

A widely understood fact is that the costs of manufacturing components in serial production are much lower than in unit production. As the proposed mechanism is completely different, it is not possible to easily switch production lines from the production of current engine parts to components necessary for the assembly of the barrel engine. At the stage of prototype research, the cost of such a transition was not estimated. It is a subject for future work, when the application of barrel engines is determined and prototype tests for specific customer groups are carried out. Then, a simpler engine design for series production will be developed and production costs can be reliably determined. It should be remembered that thanks to a large number of axisymmetric parts and a reduction in the number of important, complex components (cylinder head, camshaft with the valvetrain, crankshaft), production can be relatively profitable. Much more parts can be manufactured by turning than by milling.

Another challenge with unit production of opposed piston engines is the need to inject fuel perpendicularly to the cylinder axis. It would be ideal to have the possibility to optimize the spacing and diameter of the holes in the injector, but the unit production of such a complex element is also unprofitable. In high-volume production, this should not be a problem.

Another important issue in the context of the discussed application is the achievable scale of production of these engines. In order to notice a clear improvement in the overall vehicles' emission, it would be necessary to convince most drivers to buy such an unconventional design. This would mean that the approach we propose would have to be recognized around the world. The decision to introduce wobble plate engines to the automotive industry would have to be associated with an update of the car structure to fit an engine with a new shape. There would be no need to develop new road infrastructure, as is the case with electric cars, where the rapid development of vehicle charging stations would be necessary. Instead, it would be essential to invest in training of mechanics who have not encountered wobble plate engines in their work so far.

A popular opinion regarding two-stroke engines is the excessive use of oil and its combustion, which prevents such engines from meeting current emission standards. This issue is very well discussed in [13] where Achates Power shows that their opposed piston two stroke (OP2S) engine achieves oil consumption levels similar to a four-stroke (4S) engine. Tests on PAMAR engines confirm this observation, as during the tests no excessive oil consumption by the engine was observed. This is because the uniflow scavenged engine does not mix the lubricating oil with the fuel. Such an engine has two sets of rings-oil and compression rings, which work in the same way as in a four-stroke engine. Another frequently mentioned disadvantage of two-stroke engines is the wear of the piston rings, which occurs due to their contact with the intake or exhaust ports. Work with the PAMAR engine showed that the proper finishing of the ports' edges is enough to prevent ring failure. It's also worth noting that the oil rings do not come into contact with the ports.

2. Advantages of the PAMAR Engines

The further discussion of the advantages of PAMAR engines is based on the test results of previous prototypes. So far, four engines have been developed, and the fifth is currently under construction. Extensive tests were carried out on the most advanced PAMAR 3 and PAMAR 4 engines, the basic parameters and capabilities of which are collected in Table 1. The engines are presented in Figure 2.

Table 1. Information summary about the most advanced wobble plate axial engines built at the Warsaw University of Technology.

Engine Details	PAMAR 3	PAMAR 4
Year of Manufacture	2009	2016
Engine Displacement (cm^3)	3000	1792
Cylinder Number	6	2
Nominal power (kW)	340 (3000 rpm)	157 (3000 rpm) [2]
Torque (Nm)	1200 (1400 rpm)	720 (700–1500 rpm)
Mechanism used	Wobble plate blocked by bevel gear	Wobble plate blocked by two sliders
Ignition type	SI/CI/HCCI	SI/CI/HCCI
Fuel	Gasoline, diesel fuel, propane, LNG	Gasoline, diesel fuel (already tested) *Low calorific gas fuels, biogas (possible)*
Additional systems	Variable Port Area [1] and Variable Compression Ratio modification possible after engine shutdown Variable Angle Shift modification possible during the engine's work	Variable Compression Ratio, Variable Angle Shift and Variable Port Area possible during engine's work
Results and main conclusions	Achieved stable operation in HCCI mode at full load [14]. Very high-power density: 0.56 kg/kW	Correct operation of all additional variability systems confirmed The engine is currently under investigation

[1] The variability systems and their operation will be discussed in later sections; [2] Nominal power and torque values for PAMAR 4 engine are estimated from the calculations.

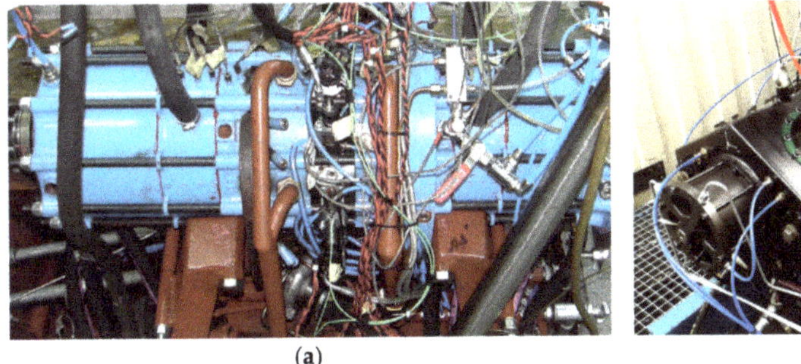

(a)

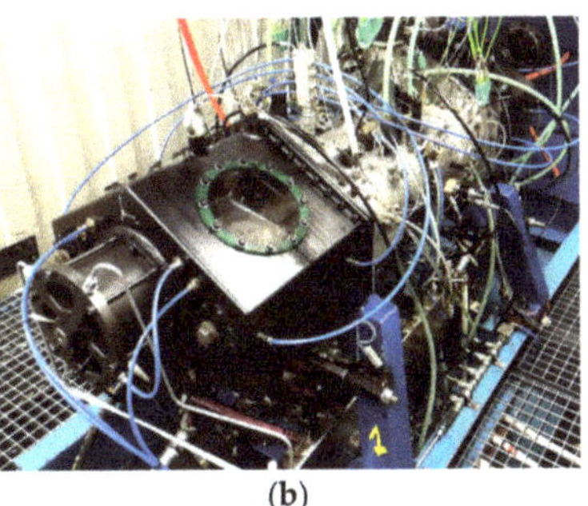

(b)

Figure 2. Two latest PAMAR engines built at Warsaw University of Technology: (**a**) PAMAR 3 (2004); (**b**) PAMAR 4 (2016).

At Warsaw University of Technology, the research on barrel engines began in 2000 with the construction of the PAMAR 1—a 50 cm^3 SI engine with wobble plate blocked by piston. The problems that arose during the preliminary tests gave direction to the further development of the analyzes, especially the need to study the course of the piston side force for individual mechanisms. The second 600 cm^3 engine was the first in the series to use an opposed piston layout as well as Variable Angle Shift and Variable Port Area possible to modify after engine shutdown. The PAMAR 2 design was based on a complete mechanical analysis performed in [3], resulting in the selection of the crosshead mechanism for wobble plate rotation blocking.

Parallel to the research on the PAMAR 4 engine, the PAMAR 5 is being built, which is to be an engine for distributed energy generation. Thanks to the additional fourth variability system, it will be possible to control the delivery of gaseous fuel to a large extent. It will enable operation on fuels of variable composition and low calorific value. The engine is designed for the power of 400 kW for fuels with a calorific value of about 6 MJ/m^3, up to 2MW for diesel. The first tests are planned for 2021. It would be a great success to confirm the engine operation on alternative fuels, as it can be an interesting research direction to reduce emissions. A review on this matter is presented in [2].

Based on many years of research, the authors have identified the most promising direction for the development of barrel engines as distributed energy sources as well as military vehicles and drones. However, in the face of clearly changing trends on the car engine market, it was decided to show the advantages of the designed engines in the context of the automotive industry.

The innovation of PAMAR engines lies in the combination of the benefits of two ideas-wobble plate mechanisms and opposed piston layout. They will be discussed in the following subsections.

2.1. Wobble Plate Benefits

In barrel engines, the sliding motion of the pistons can be converted into the rotary motion of the shaft in various ways. This is usually done using a special plate. There are two types of plates—swash and wobble plate. It is very important to distinguish between these two types, as calculations have shown that wobble plate mechanisms have the potential to achieve much greater mechanical efficiency than swash plates. It should also be taken into account that there are many varieties of wobble plate mechanisms. It is necessary to carefully analyze the dynamics of these mechanisms in order to decide whether to choose them for the design. It is impossible to talk about the advantages and disadvantages of barrel engines without fully understanding how they differ from each other.

Moreover, the lack of a precise kinematic analysis of the mechanisms may lead to failures or completely prevent operation. Such a situation took place in the frequently quoted Bristol engine [4], where the plate movement did not meet the uniform precession condition. The publication [5] noted that among the mechanisms patented in the US, some of them could not operate on the basis of the analysis of degrees of freedom. According to [5] the most representative group among the mechanisms appearing in the US patent database is the wobble plate blocked by bevel gear.

2.1.1. Lower Piston Side Force

The dominant source of engine rubbing friction is the piston assembly [15]. The reduction of the piston side force can significantly increase the mechanical efficiency. It is worth mentioning that in order to limit the frictional work generated by the side piston force, engineers decide on various modifications. An example is the shift of the cylinder axis from the axis of the crankshaft in the newest family of Mercedes engines OM 654. Importantly, the only compelling competition for engines with a crank mechanism are the engines with lower mechanical losses. Therefore, after analyzing the kinematics of the mechanism, the course of this force should be estimated. The calculation results for the above-mentioned mechanisms are shown in Figure 3. The maximum force in the classic crank mechanism (red dashed line) was taken as the reference value (100%). Significantly, in some mechanisms this force can be much greater (swashplate, cam mechanism, wobble plate blocked by piston). This may be the reason for the popular belief that axial engines are not worth considering because of their low efficiency. It should be noted, however, that there are mechanisms that are very competitive with the classical construction. These are mainly designs based on bevel gears or crosshead. The lateral piston force can be as much as 50% lower here than in a conventional engine. This is one of the reasons why these two types of engines were selected for use in the actual test prototypes.

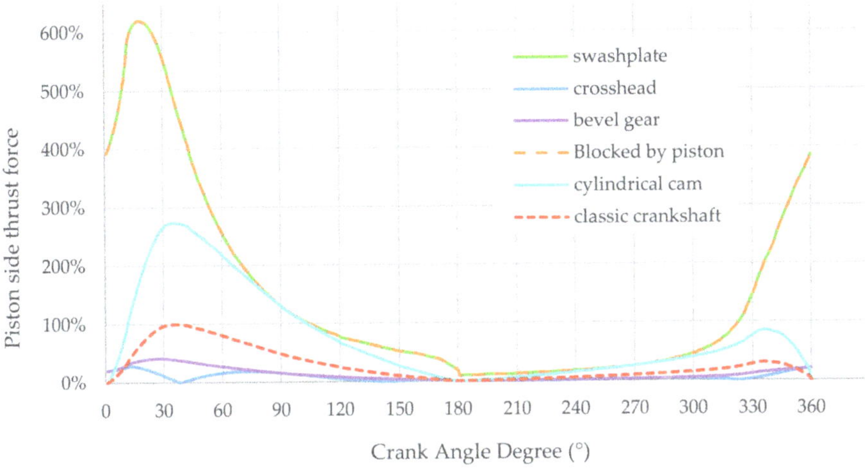

Figure 3. The course of the absolute value of different mechanisms' piston side force for the same operating parameters [3].

2.1.2. Piston and Connecting Rod Bearings

The kinematics of the wobble plate mechanism requires the use of spherical plain bearings in the connecting rod-piston and connecting rod-wobble plate joints (Figure 4). This type of bearing is characterized by a much higher load capacity than the configuration in which the force is transmitted through ordinary slide bearings and the piston pin. An example of the use of a spherical bearing in an internal combustion engine was provided by the Sulzer company, which patented the rotating piston in 1937. After a long period of testing, it was first used in 1964 in Z-type engines, and in 1995 the idea was adopted by GMT (Grandi Motori Trieste) in the VA55.

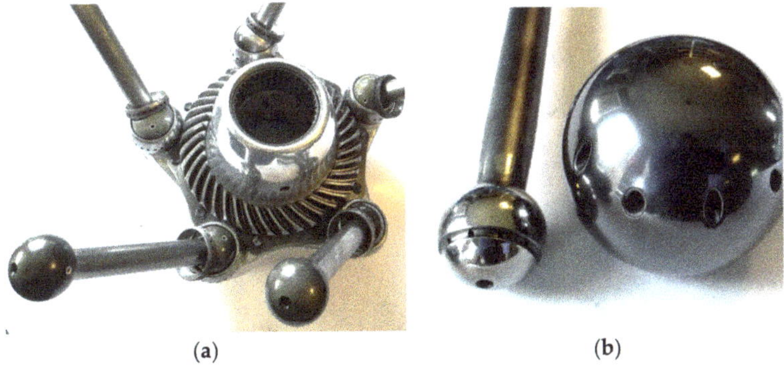

(a) **(b)**

Figure 4. High capacity spherical plain bearings used in PAMAR engine designs: (**a**) wobble plate—connecting rod assembly of PAMAR 3 engine; (**b**) size comparison between the bearings used in PAMAR 4 and PAMAR 5 engines.

Experience has shown that this solution has the following advantages [16]:

- 40% more bearing area than a gudgeon pin bearing.
- More even distribution of load and temperature in axis symmetrical pistons.
- Symmetrical deformations of the piston which allow smallest running clearance between piston and cylinder liner.
- Possible use of the rotation mechanism, which will ensure lower wear of the piston.

In PAMAR engines, no additional piston rotation mechanism was used. However, during the inspection of parts, after some time of engine operation, spiral lines were observed on the piston skirt surface. It proves the existence of low revolutions of the piston during engine operation and may be a result of the velocity components of the connecting rod that do not appear in the classical mechanism.

2.1.3. Additional Engine Control Systems

The ability to adjust the engine operation to varying loads has become an important issue in the automotive industry. As summarized in [17], research was conducted towards systems such as variable valve timing, variable length intake manifold, variable spark timing and variable compression ratio. The use of these systems was expected to have a beneficial effect on engine performance along with reduced fuel consumption and lower CO_2 and NO_x emissions. However, one of the main design goals is to make as few changes to the engine as possible, preferably to only slightly modify a stock engine, to ensure reduced introduction to market time.

In barrel engines, the implementation of such additional mechanical control systems is relatively simple. The design of the systems in PAMAR engines is extremely flexible due to the combination of three assumptions: the wobble plate mechanism, opposed piston layout and two stroke work cycle. As also noted in [5], systems of variation would not be so simple to implement on the basis of classical crank and connecting rod assembly. This has been confirmed in the research of PAMAR engines and the simplicity of the implementation of these systems is one of their crucial advantages. Already tested system designs will be the subject of patents filed in the near future, therefore their detailed operating principle is out of scope of this paper. The systems used in the PAMAR 4 engine are: Variable Compression Ratio, Variable Angle Shift and Variable Port Area.

Variable Compression Ratio

The Variable Compression Ratio system has been under engineers' consideration since the beginning of the 20th century. The results of their work could be divided into following categories [18]: unconventional cranktrains (NISSAN), systems with variable head-crankshaft distance (SAAB) and variable kinematic lengths (FEV conrod).

Various approaches have been described to alter the compression ratio and their advantages as well as disadvantages have been examined. The system used in PAMAR engines will not be described in detail, as it will be the subject of a patent in the next few years. However, assessing it on the basis of the parameters proposed in [17], it can be certainly stated that in opposed piston wobble plate engines it is possible to implement a VCR system, with:

- preserved combustion chamber integrity and engine overall rigidity,
- no changes in crankshaft-piston assembly kinematics,
- no significant influence on mechanical losses due to the possibility of using a well-lubricated tribological pairs,
- no variations on engine displacement,
- very good control accuracy. The compression ratio may vary smoothly from 1:10 to 1:20 while the engine is running.

Variable Angle Shift and Variable Port Area

Other crucial systems widely researched in the automotive engines, namely Variable Valve Timing and Variable Valve Lift, are also implemented in the PAMAR engine. Due to the lack of classic valves in the barrel engine, these systems are named Variable Angle Shift and Variable Port Area, respectively. The principle of operation of these two systems is presented inFigure 5.

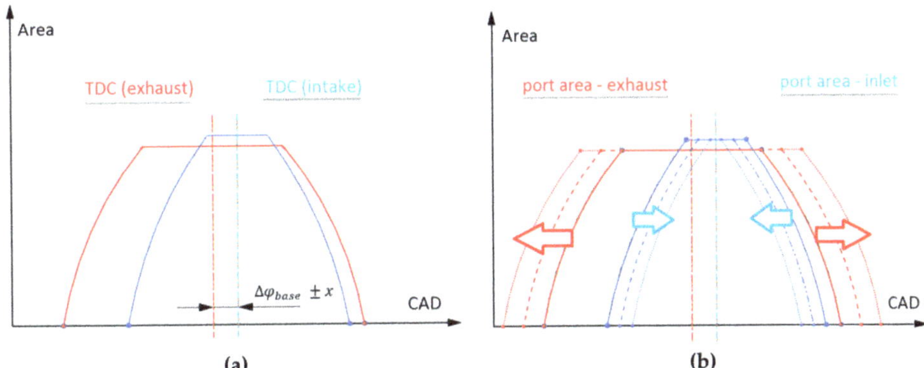

Figure 5. Time-area plots for visualization of the idea of two variability systems implemented in PAMAR engines: (**a**) Variable Angle Shift; (**b**) Variable Port Area.

Variable Angle Shift is responsible for the shift in the opening phase of the intake port relative to the exhaust port. This is an equivalent to a phase shift of the two pistons that control these ports. The shift can take place continuously while the engine is running in the range of ±8 degrees from the $\Delta\varphi_{base}$ value (in the PAMAR 4 engine). The base value can be changed in any range after dismantling the engine.

Variable Port Area changes the opening angle of individual ports. Increasing the intake port cross-section reduces the exhaust port cross-section. Changes can be made while the engine is running.

The systems are generating considerable interest due to the flexibility they provide. They allow among other things, for the control of the HCCI combustion process, as stated in review on this matter [19]. Manufacturers try to use their potential despite the necessity to significantly complicate the structure. It becomes profitable for them, especially in the face of increasingly strict emission standards. As concluded in [20], the use of VVT can be associated with a significant reduction in NO_X, CO and HC emissions. It also allows the range of engine operation to be extended to lower engine speeds, further increasing engine efficiency. The research also shows torque gains in engines using VVT, which are mainly the result of lower pumping losses and increased volumetric efficiency.

Two-stroke engines require optimal control of the opening and closing angle of the intake and exhaust valves for proper scavenging efficiency. Using the system allows to control the internal Exhaust Gas Recirculation (EGR) amount.

Particular attention should be paid to the control of port opening in an opposed piston two-stroke engine, because the increase in phase shift between the pistons covering the ports causes a decrease in mechanical efficiency. This is because both pistons have a single combustion chamber but are out of phase resulting in an uneven ability to generate torque on the shaft [21].

The systems discussed above can also have an impact on the aftertreatment emission control. VVT is mentioned as one of the catalytic converters' thermal management methods during cold start and warm up [22]. Through the implementation of late intake valve opening and early exhaust valve opening strategy it is possible to increase exhaust temperature and reduce catalyst light-off time. However, such approach leads to a decrease in brake thermal efficiency.

2.1.4. Barrel Shape and Axial Symmetry

The characteristic barrel shape of the axial engine (Figure 6) has also its advantages. One of them-especially important in aviation-is the small frontal area. From a thermodynamic point of view, axial symmetry can enable the same filling, lubrication and cooling conditions for all cylinders. This makes the process of controlling optimal engine operation easier. An important aspect also discussed when assessing the variability systems described in the previous subsection is the ability to control them in all cylinders. This problem does not occur in barrel engines, where the conditions in all cylinders are the same by default. It is a desired feature, especially considering the implementation of sophisticated combustion strategies as HCCI, which is a subject of interest of engine manufacturers due to its low NO_x and particulate matter (PM) emissions [19].

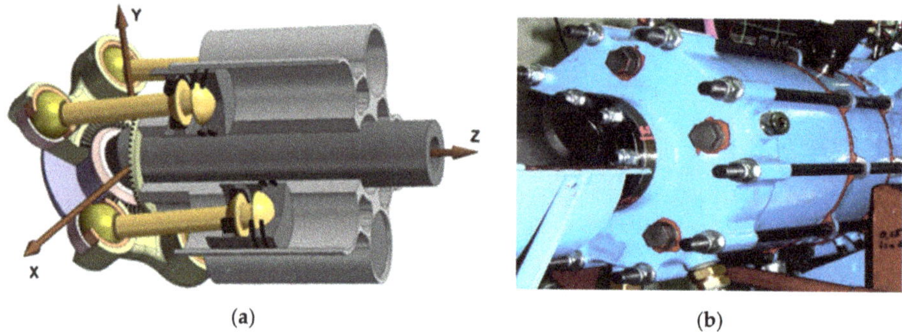

Figure 6. The characteristic shape of an axial engine: (**a**) a simple CAD model of wobble-plate crank assembly; (**b**) The axissymmetrical shape of the PAMAR 3 engine.

2.1.5. Torsional Vibrations

Time-varying gas forces and inertia forces act on the crankshaft, causing it to vibrate. There are three types of vibrations in the crankshaft: longitudinal, bending and torsional, the latter of which are the most dangerous. The torsional deflection is not limited by anything other than the stiffness of the shaft, which can lead to high vibration amplitudes. This may cause an increase in fatigue stress and structure failure. For this reason, a vibration calculation of the crankshaft is necessary, with particular attention to the natural frequency, to prevent resonance. In order to reduce the harmful effects of this phenomenon, vibration dampers of various designs are used.

The problem of torsional vibrations is a big challenge for designers of opposed piston engines, especially engines with two crankshafts connected by a gear train. If the crank angle offset system is used—similarly to the previously discussed Variable Angle Shift, the load is not evenly distributed over both shafts. This can be seen in historical designs of opposed piston engines, using dual vibration dampers (Leyland L60) on one of the shafts or using one of the shafts only to drive the accessories, to limit the flow of variable torque through the transmission connecting the shafts (Junkers) [21].

In an engine with a wobble plate mechanism, the problem of torsional vibrations is significantly reduced. The gas and inertia forces are first transferred to the rigid wobble plate and then transmitted to the shaft. It is worth noting that due to the significant difference in the method of transmission of the drive in the wobble plate mechanism, the shaft has a much higher torsional stiffness than the classic crankshaft.Figure 7shows a schematic cross-section of the PAMAR 4 engine with the rigid components responsible for transmitting the torque marked in green.

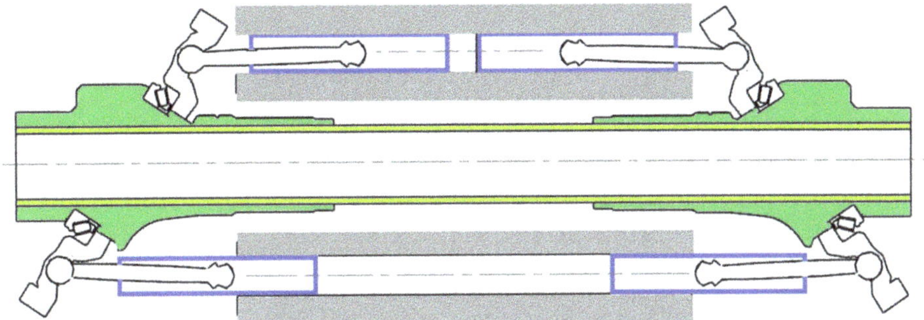

Figure 7. The wobble plate mechanism of PAMAR 4 engine.

2.2. Opposed Piston Benefits

The arrangement of opposed piston layout began to attract the interest of designers and investors, who want to adapt to the restrictive requirements faced by modern engines. The largest company promoting this type of engine is Achates Power, which has already been very successful in using opposed piston engines in a variety of industries—from passenger cars to military vehicles [23]. A massive investment has also been made in the OPOC engine—opposed piston opposed cylinder-key financial backers are Vinod Khosla and Bill Gates [24].

In the recent years there has been growing interest in opposed-piston engines, particularly the two-stroke engines. The advantages of such a solution have been described in detail in many publications, including [25,26] and [14]—especially concerning PAMAR engines built at the Warsaw University of Technology. The recurring interest in opposed piston engines is due to the numerous advantages that are important in view of the increasingly stringent emission standards. The old problems that pushed this idea aside can be solved with modern technology, and the efficiency gain is worth the adaptation to the completely new cylinder layout. As confirmed by theoretical analyzes [25], the most advantageous system in terms of configuration is the opposed piston engine operating in a two-stroke cycle.

In [25] comparison was made between three types of engines: classic four-stroke (4S), four-stroke opposed piston (OP4S) and two-stroke opposed piston (OP2S) with the closest geometric characteristics possible. For all the engines indicated power, engine speed, and maximum pressure rise rate were assumed constant. The analysis showed clear benefits of the opposed piston system, calculated as significantly lower energy losses related to the parameters determining the differences in the actual course of the cycle and the ideal Carnot cycle. These parameters are heat transfer, finite duration of the combustion and variable specific heats of the working fluid. A particularly large gain was seen in the two-stroke engine in parameter related to the combustion duration, which results from the doubled firing frequency. More detailed calculations showed approximately 10% lower fuel consumption in comparison to the four-stroke opposed piston engine. The possible reasons of these advantages, widely described in literature, are briefly discussed below.

2.2.1. Less Heat Rejection to the Cooling System

The internal combustion engine has a strictly defined efficiency limit which is caused by the second law of thermodynamics. Attempts to approach this border consist mainly in modifications aimed at limiting the heat exchange between the working medium and its surroundings, i.e., mainly piston crown, cylinder head, poppet valves and cylinder liner. Designers are limited by the strength of materials at elevated temperatures. Taking into account that the temperature peak during the combustion process is up to 2500 K, cooling the most thermally loaded parts of the engine is necessary. The heat rejected to the cooling system is a waste of energy which, to some extent, could be converted into useful work performed by the engine.

The engineers were able to use continuously better materials and insulating coatings, but cooling parts such as the cylinder head or poppet valves was still a necessity and contributed to heat loss. Opposed piston layout might be the answer to this problem. The engine timing based on the piston movement means that it is not necessary to cool or drive additional poppet valves. A great advantage is also the lack of classic cylinder head.

The advantage described above is apparent from the historical opposed piston engine data.Figure 8 shows the heat loss to the cooling system of opposed piston engines in comparison to the classical engine designs. The clearly visible advantage of 5–10 percentage points of fuel energy is a result of the lower amount of parts requiring cooling and the favorable Volume to Area ratio (V/A) ratio discussed in the next subsection.

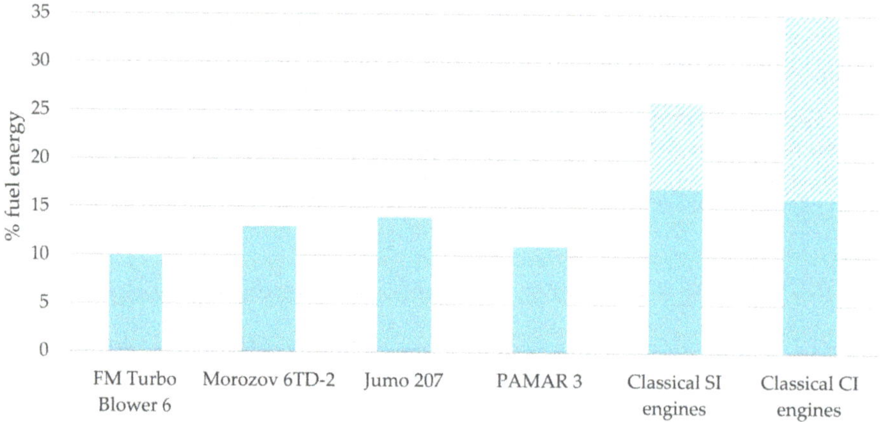

Figure 8. Heat loss to the cooling system of historical opposed piston engines [18] compared with the classical SI and CI designs [15]. The comparison applies to the CI OP2S engines: FM Turbo Blower 6 (1938)–102 dm^3; Morozov 6TD-2 (1967)–16,3 dm^3; Jumo 207 (1939)–16,6 dm^3; PAMAR 3 (2006)–3 dm^3.

Differences for individual engines may of course be caused by different swept volume and operating conditions, as engines with a larger capacity will have lower heat losses. The year of production of the engine is also important, as the current engines have lower heat losses due to the use of various design improvements. In Figure 8, the shaded area for classic constructions represents the range of the described parameter given by Heywood [15], who refers to the research from 1957–1974. For reference, the engine capacity and year of manufacture are stated for each engine presented in the figure.

2.2.2. Stroke to Bore Ratio

When designing an engine, the Stroke to Bore ratio (S/B) is decided based on several criteria. Taking into account the desire to reduce the heat exchange, the most advantageous shape of the combustion chamber will be characterized by the largest possible volume and the smallest area. From a mathematical point of view, the geometric shape with the greatest V/A ratio is the sphere. Hence the conclusion that the most advantageous proportions of the combustion chamber must be close to the shape of the sphere. Considering the dimensions of the combustion chamber, when the piston is in the Top Dead Center (TDC), such a system is obtained using the greatest possible stroke to bore values.

Another argument in favor of high S/B is the higher efficiency of uniflow scavenging when the air has a longer distance to travel between the intake and exhaust ports. It prevents a phenomenon known as short circuiting—loss of fresh charge through exhaust ports.

The research on HCCI shows that a higher stroke to bore ratio means greater stratification of the charge, which lowers the pressure rise rate. As a result, the engine has a lower knocking tendency.

Calculation of the optimal S/B ratio is out of scope of this paper, but it should be borne in mind that the values of this parameter obtained in currently produced engines are not the result of thermodynamic optimization alone. The upper value of this parameter is limited, among others, due to the limit of the allowed piston speed and the collision of the connecting rod with the cylinder surface.

The opposed piston arrangement may be the answer to some of these limitations, as here two pistons move in one cylinder. As a result, the stroke value is double the displacement value of a single piston. Thus, an increase in the S/B ratio does not increase the linear speed of the piston. Approximate values of this parameter for the opposed piston crankshaft engine can be estimated on the basis of data from Achates Power, with stroke to bore ratio is within 2.2–2.6 [27].

The second limitation is the size limitation, which is the result of a possible collision of the connecting rod with the cylinder. For this reason, the highest S/B values are characteristic for low-speed marine engines, where a collision cannot occur because of the crosshead mechanism used. As reported by [16], the parameter value for Sulzer marine engines in the late 1970s was around 2 (RLB engine) and has doubled by the early 1990s, continuously growing (Wartsila X35-4.43). The change in this parameter over the years is presented in Figure 9.

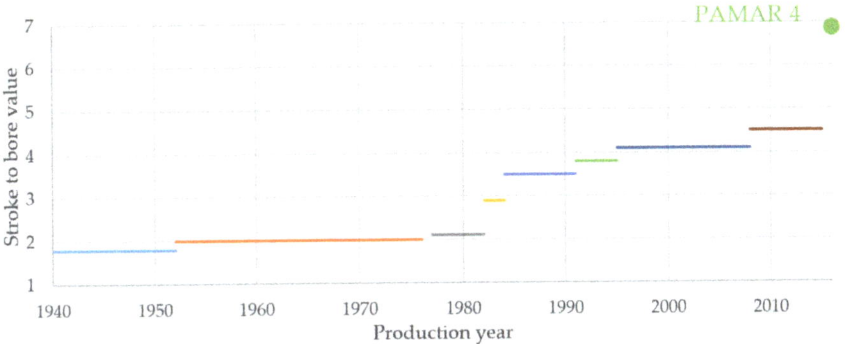

Figure 9. Stroke to bore values for low speed marine engines (adapted from [28]). PAMAR 4 data for comparison.

It is worth noting that the size limitation is less significant in wobble plate engines than in classic engines with a crank mechanism, because the connecting rod deviation angles from the cylinder axis are smaller. This is expressed in the lower value of the piston side force discussed earlier and in the possibility of obtaining higher S/B ratios. Thanks to this, the PAMAR 4 engine managed to achieve a stroke to bore ratio of 6.85 at a mean piston speed of about 9.4 m/s at a nominal rotational speed of 1500 rpm. The stroke to bore proportions of the PAMAR 4 engine are clearly visible in Figure 7.

2.2.3. Uniflow Scavenging

The possibility of controlling the opening of the intake and exhaust ports by opposed pistons creates ideal conditions for uniflow scavenging. This type of scavenging is characterized by the greatest efficiency [15].

As shown by the calculations [29], there is also a relationship between the scavenging efficiency and the previously mentioned stroke to bore ratio. For the four S/B geometries analyzed there in the range from 1.08 to 2.5, it turns out that the highest parameter value corresponds to almost ideal scavenging conditions, with the efficiency increasing from 0.813 to 0.934. The analysis of the distribution of velocity vectors shows that with an increase in S/B it is possible to obtain uniform flow field with no recirculation zones. Due to the shorter axial distance between the inlet ports and the cylinder axis, the air has time to form a uniform front which pushes residual gas out of the cylinder more efficiently.

2.3. Downsizing and Downspeeding as PAMAR Engine Potential Demonstration

All of the above features of barrel engines mean that its performance can be significantly higher than in the case of classic designs. In order to support this thesis experimentally, the research of the PAMAR 4 engine conducted at low rotational speed was presented in the further part of this paper. The tests were aimed at determining the torque curve for the low speed range and presenting the engine performance in the context of two trends widely present in the automotive industry, i.e., downsizing and downspeeding.

Downspeeding is a procedure aimed at shifting the engine's operating point towards lower rotational speed. The advantage of such action is the increase in mechanical efficiency and reduction of pumping losses. At lower speeds, the coefficient of friction decreases. The piston will perform fewer work cycles, which results in less friction work. Filling the cylinder is more efficient by lowering the flow velocity, which reduces the time frame when the inlet velocity is close to the speed of sound.

As presented in the publication [30], in commercial trucks for highway cruise speeds, the decrease in engine speed by every 100 rpm translates into one percentage point of fuel efficiency increase. Carbon dioxide emissions reduction was estimated at 12,000 lbs (5443 kg) per truck annually.

Downsizing is one of the most discussed design improvement directions when it comes to the review of technology progress concerning the introduction of greenhouse gas emissions and fuel economy standards. As the International Council of Clean Transportation estimates, the market share of downsized engines will be increasing significantly in the coming years [31]. The turbocharged vehicle sales have increased from 3.3% in 2004 to about 20% in 2015. A study prepared by ICCT concluded that the main advantage of downsized engines is low incremental cost compared to their fuel efficiency benefit.

Downsizing requires designers to allow higher power density, ensuring reliability achieved in larger units of the same power. Working with a higher Break Mean Effective Pressure (BMEP) allows to achieve lower throttling losses under normal driving conditions. Smaller displacement engine is likely to have smaller friction losses and lower weight.

The trend of downsizing is related to research on the most effective turbocharging methods, where the most interesting concepts include e-boosting and Variable Geometry Turbocharger. The advantage of turbochargers is the partial energy recovery of exhaust gases.

The variable compression ratio is advantageous for downsized engines. This system will ensure that knocking combustion is controlled at a higher load, while allowing for greater efficiency under normal driving conditions.

3. Experimental Setup for Torque Curve Calculation

The research was conducted on the PAMAR 4—the last in a series of engines built at the Warsaw University of Technology. Its parameters are presented in Table 2 in juxtaposition with the data of two Mercedes engines, which will be used to compare the performance of the three power units. The additional information on the systems used in the axial engine under consideration is provided in the Table 1 at the beginning of this article. The engine ran in the CI mode; however, the SI is also possible after changing the compression ratio and slightly modifying the experimental setup.

Table 2. The PAMAR 4 engine parameters compared to the Mercedes engines [32,33].

Engine Details	OM651 DE18	OM654 DE16	PAMAR 4
Engine displacement (cm^3)	1796	1598	1792
Bore (mm)	83	78	55
Stroke (mm)	83	83.6	188.6
Stroke/bore ratio (–)	1.0	1.07	2 × 3.43 = 6.86
Nominal power (kW) (at speed (rpm))	100 (3400–4400)	118 (3800)	157 (3000)
Max torque (Nm) (at speed (rpm))	300 (1600–3000)	360 (1600–2600)	720 (700–1500)
Compression ratio	16.2	15.5	10–20

The experiments were aimed at testing the engine operation at low rotational speed. The parameters changed during the tests included: compression ratio (VCR system), time-area plots of the inlet and exhaust ports and their overlap (VAS and VPA systems), fuel dose, number of injections and their timing, charging pressure, Variable Turbine Geometry VTG (turbine blades angles). The article presents ten series of tests that allowed for the analysis of engine operation for various values of the parameters mentioned above. Details of each test series are presented in Table 3. During each series, based on the observation of engine operation, the control values were changed in the ranges given in the table. The goal was to achieve stable operation and the highest possible torque.

Table 3. Experiment details for ten test series presented in the article

Test Series	Compression Ratio	Angle Shift	Port Area [1]	Inlet Manifold Absolute Pressure (bar)	Number of Injections	Injection Angles (°) [2]	VTG Angle (%) [3]	Maximal Torque [4]	Speed Range (rpm)
A	19.8–17.3	9.8–17.5	0–4.8	1.4–2.6	3	−2.1/−3.1	70–100	377	410–610
B	18.1–14.8	9.1–15.5	0–4.3	1.7–3.1	4	−1.8/+1.8	60–95	363	410–565
C	17.5–15.6	8.8–14.6	0–5.5	1.5–2.8	3	−17.3/+1.4	70–90	337	440–560
D	16.8–14.8	8.2–12.6	0–2.8	1.8–3.4	5	−20.8/+4.1	65–100	407	320–450
E	17.2–14.5	10.2–18.1	0–6.8	1.6–3.6	2	−25.7/+3.1	30–70	414	420–660
F	16.1–13.4	8.8–16.3	0–4.2	2.2–4.1	2	−20.3/+4.2	40–80	503	430–530
G	17.7–15.3	8.2–14.7	0–5.1	1.4–3.2	3	−21.2/+2.4	60–100	427	400–570
H	16.3–13.9	9.3–13.9	0–6.1	1.8–3.6	2	−16.8/+1.8	50–90	466	340–450
I	15.5–14.1	9.9–17.2	0–6.4	1.5–3.7	3	−19.5/+0.5	45–100	470	500–570
J	18.2–16.1	8.5–16.5	0–4.8	1.4–2.8	4	−16.2/+1.5	65–95	402	465–520

[1] Port area parameter is possible to set in the range of 0–8. A zero value of parameter corresponds to the base value of exhaust port area. Parameter equal to 8 means increasing exhaust port area by approximately 20%. [2] For each series, a range of angles represents the period from the start of the first injection to the end of the last injection. 0 °Corresponds with the TDC (in OP engines defined as the angle of minimal volume between two cylinders) [3] The range from 0 to 100% represents the ratio of turbine blade angle in the range made possible by the turbocharger manufacturer. 0% means the maximum width of the slots and the minimum inflow velocity to the turbine. [4] The maximal torque value obtained from averaging over 100 cycles.

The article focuses on presenting the relationship between torque and rotational speed. Additionally, pressure in the cylinders and pulsations in the intake and exhaust channels were measured. The acquisition system has the possibility of monitoring temperatures along the cylinder wall and at the piston crown, as well as pressures and temperatures at all characteristic points in the installation. The basic elements of the test stand are shown in Figure 10.

(a)

Torque	Froude Hoffman AG400HS eddy current dynamometer
Engine speed	Heidenhein ERN 130 encoder
Cylinder pressure	IMES FPS-01 Kistler 6055 C80
Injector	Bosch 0 261 500 172
Turbocharging	Garret Variable Nozzle Turbine GT1749V
Supercharging	Eaton TVS R410
Data acquisition	NI PXI system with customized modules
Engine control software	Customized LabVIEW code

(b)

Figure 10. Experimental setup: (**a**) the PAMAR 4 engine on the testbed; (**b**) testing equipment used during experiment.

4. Results and Discussion

Figure 11 shows the results of the engine tests for the different experiment series. Each point shows specific parameters of rotational speed and torque at which the engine has worked. Bigger red dots in the figure represent several distinctive experimental points achieved at a stable operating condition of the engine that are characterized by a minimum rotational speed and maximum torque. On the basis of selected reference points, a linear trend function was determined and plotted as the red line in the Figure 11. As for preliminary research, the calculated function is a good approximation of the torque curve for low rotational speeds. A 100-cycle averaged indicator diagram is presented in the Figure 12. It is a diagram for the F test series, where the torque of 500 Nm was achieved. The maximum pressure in the cylinder is 185 bar, and the cycle work is 1944 J.

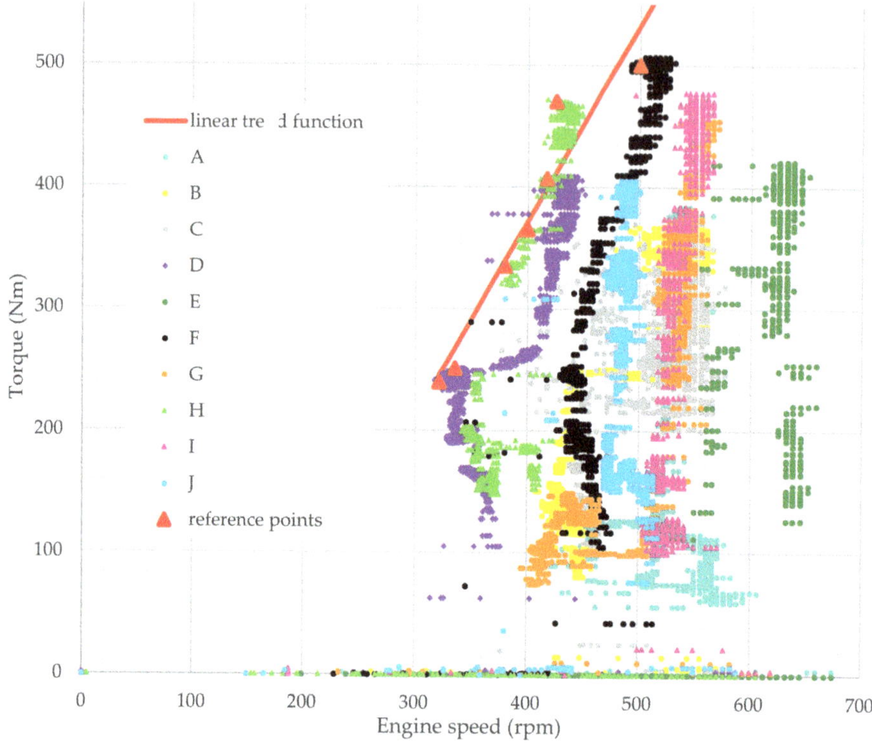

Figure 11. Engine operating points achieved during tests.

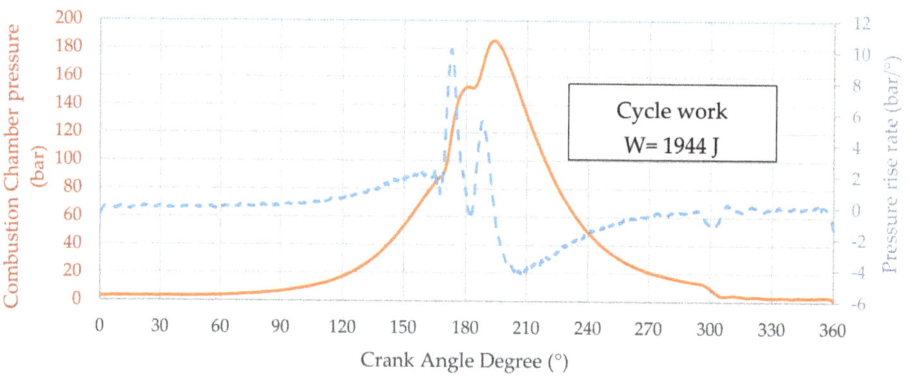

Figure 12. Averaged indicator diagram for the maximal torque in F test series. The maximal pressure rise rate was about 10 bar/°.

The conducted research allowed to update the predicted engine parameters, as shown in the Figure 13. The red points on the graph correspond to the reference points in Figure 11. On their basis, the updated torque curve was determined, which is shifted by about 80 Nm compared to the design prediction. At the design stage, assumptions were made regarding some key parameters, which allowed to estimate the maximum torque expected in the PAMAR 4 engine. The maximum allowed pressure in the combustion chamber was assumed to be 300 bar. The preliminary assumptions included the maximum compression ratio of 1:20, the maximum absolute boost pressure of 4.5 bar provided by both the supercharger and turbocharger. Constant volume heat addition pressure ratio was estimated to be around 3, and the maximal pressure was expected to appear 5° before TDC. The heat flux lost from the combustion chamber has been limited by the special design of the piston, allowing the maximum temperature at its crown at 700 °C and for the combustion chamber walls at 600 °C. Basing on these parameters (solid green line on the graph), the stable operation of the engine was to be achieved at a speed of about 400 rpm and reached the nominal value of 720 Nm at 800 rpm. The experiment has shown that these expectations can be raised, as shown by the dashed green line in the figure. The torque of 240 Nm was successfully achieved at the rotational speed of around 300 rpm, which provides powerful evidence for the engine's downspeeding capability. Taking into account the maximum torque of 500 Nm obtained during the experiment with as much as 180 bar, it was found that with a very high probability the result of 800 Nm can be achieved without exceeding the maximum allowable pressure in the cylinder.

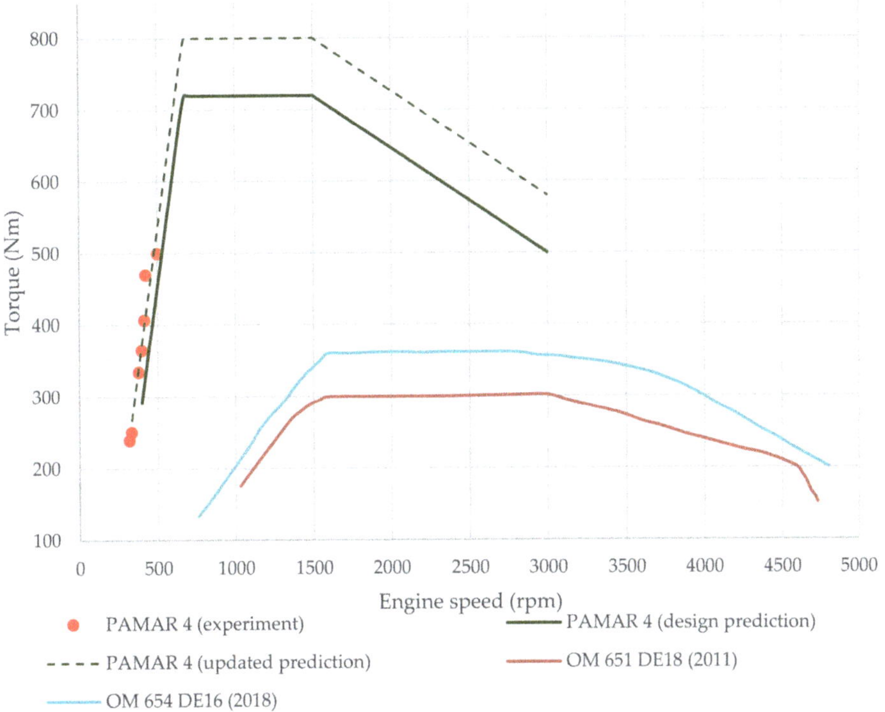

Figure 13. Torque-speed curves of the PAMAR 4 engine. The characteristics of the tested engine for the range of 500–3000 rpm comes from theoretical calculations. The results are compared with the data from OM651 engine [32] and OM654 (estimated from [34]).

The published results are primarily intended to show quantitatively how the design advantages of PAMAR engines described in the article affect their performance. In order to refer the results obtained on the testbed to the state of the art, they were compared with the torque curve provided by Mercedes for the OM 651 DE 18 engine. This engine was selected due to its similar capacity, the same type of ignition and good reputation of the manufacturer.

When comparing the two power units, it should be noted that the OM651 DE18 engine was released in 2011, five years earlier than the PAMAR 4. Its successor, the OM 654 engine family, was launched in 2016. It includes, among others unit OM 654 DE16 with a capacity of 1597 cm^3, which develops a maximum torque of 360 Nm. For reference, the most powerful engine in the series, the 1950 cm^3 OM 654 DE20 achieves a record 500 Nm of torque and 180 kW of power [33].

The difference in the torque curve as a function of rotational speed, visible in the graphs, cannot be explained by the time interval only. The high torque value at low rotational speed is the result of the advantage that the design of the barrel engine has over classic engines. The most important role here was played by the possibility of smooth control of the variability systems, which is not available in the series of Mercedes engines presented in the paper.

Looking at the torque diagrams for two engines released over several years, one can clearly see the downsizing trend towards achieving the highest possible performance from the smallest capacity unit. There is also an intention to shift the torque curve towards lower rotational speeds. PAMAR engines are in line with these trends, offering torque that has not yet been achieved in any other car engine of similar capacity.

5. Conclusions and Future Work

According to the authors, the type of engine presented in this paper may be an interesting direction in the development of internal combustion engines in the face of the threat of their complete withdrawal. The article proves that, thanks to a precise kinematic analysis, it is possible to build an opposed piston wobble plate engine, which will not have the disadvantages historically attributed to it. Additionally, the unconventional shape of the designed engines allows for the implementation of variability systems that are the subject of research by key engine manufacturers. These systems are: Variable Compression Ratio, Variable Angle Shift and Variable Port Area. The tests presented in the article were carried out with various settings of these systems changed during engine operation, which proves that their control has been achieved.

The article demonstrates that the PAMAR 4 engine is able to generate significantly more torque than a Mercedes power unit of the same capacity. It has been proven that the engine can run at very low speeds in the range of 300–500 rpm. The maximum torque of 500 Nm achieved during the tests does not correspond to the nominal parameters-it was achieved at a rotational speed of only 500 rpm. The maximal torque value assumed at the design stage was 720 Nm. Due to the mechanical design of the engine with the maximum pressure in the combustion chamber of 300 bar and the high permissible temperatures of the piston crown and cylinder, the test results allow the assumption that the nominal parameters will be exceeded.

The analysis of the designed engine based on theoretical calculations and the results of the research presented in the referenced publications shows that PAMAR engines have a chance to achieve competitive values of mechanical and thermodynamic efficiency.

Future work on the PAMAR 4 engine will focus, among other things, on the experimental demonstration of the high efficiency of the engine. Torque tests for higher rotational speeds will be carried out first. Then, after installing a precision fuel flow meter on the stand, it will be possible to determine the fuel consumption map. Subsequent tests will be carried out on new fuels, the first of which will be methanol and ethanol. The research will also be carried out for the implementation of the HCCI combustion, which has already been carried out in PAMAR 3. The interest in HCCI combustion is mainly due to its low NOx and PM emissions. For 2021 there are tests of the currently assembled PAMAR 5 planned, in which one of the designed systems is to be responsible for dosing gaseous fuel with different calorific values.

The advantages of barrel engines presented in the article result only from their mechanical design. This means that they can be an excellent basis for further research, guaranteeing from the very beginning a higher efficiency than engines with a classic crank mechanism. Work on further reduction of emissions can be directed-as in other engines–towards researching ecological fuels, additional injection of various additives (water, urea) into the combustion chamber and manifolds or the use of catalytic converters.

Author Contributions: Conceptualization, P.M. and B.M.; methodology, P.M., B.M.; formal analysis, B.M.; investigation, P.M.; writing—original draft preparation, B.M.; writing—review and editing, P.M.; visualization, B.M.; supervision, P.M.; project administration, P.M.; funding acquisition, P.M. All authors have read and agreed to the published version of the manuscript.

Funding: This work is a part of Applied Research Programme of the National Centre for Research and Development within the scope of applied research in industry branches (programme path B) „Badania wysokosprawnego silnika wykorzystującego technologię HCCI do zastosowań w energetyce rozproszonej" (GENEKO)-contract number PBS3/B4/16/2015.

Conflicts of Interest: The authors declare no conflict of interest.

Abbreviations

VCR	Variable Compression Ratio
SI	Spark Ignition
CI	Compression Ignition
HCCI	Homogenuous Charge Compression Ignition
CAI	Controlled Auto Ignition
ICCT	International Council of Clean Transportation
VVT	Variable Valve Timing
EGR	Exhaust Gas Recirculation
V/A	Volume-Area ratio
S/B	Stroke to Bore ratio
TDC	Top Dead Center
VAS	Variable Angle Shift
VPA	Variable Port Angle

References

1. Wappelhorst, S. The End of the Road? An Overview of Combustion-Engine Car Phase-Out Announcements across Europe. ICCT. 2020. Available online: https://theicct.org/publications/combustion-engine-car-phase-out-EU (accessed on 24 August 2020).
2. Martins, J.; Brito, F.P. Alternative Fuels for Internal Combustion Engines. *Energies* **2020**, *13*, 4086. [CrossRef]
3. Mazuro, P. Reciprocating Engines with Cylinder Axis Parallel to the Drive Shaft—Theory and Practice. Ph.D. Thesis, Warsaw University of Technology, Warsaw, Poland, 2006. (In Polish).
4. Yu, Z.; Lee, T.W. Kinematic Structural and Functional Analysis of Wobble-Plate Engines. *ASME J. Mech. Trans. Autom.* **1986**, *108*, 226–236. [CrossRef]
5. Kutenev, V.; Romanchev, Y.; Zlenko, M. *Axial Internal Combustion Engine—Practical Prospects for the Future*; SAE Technical Paper 940204; SAE International: Pittsburgh, PA, USA, 1994. [CrossRef]
6. Duke Engines. Available online: http://www.dukeengines.com (accessed on 26 August 2020).
7. Covaxe. Available online: https://www.covaxe.com (accessed on 26 August 2020).
8. Thompson, G.; Wowczuk, Z.; Smith, J. *Rotary Engines—A Concept Review*; SAE Technical Paper 2003-01-3206; SAE International: Pittsburgh, PA, USA, 2003. [CrossRef]
9. Deng, H.; Pan, C.; Wang, X.; Zhang, L.; Deng, L. Comparison of two types of twin-rotor piston engine mechanisms. *J. Cent. South Univ.* **2013**, *20*, 363–371. [CrossRef]
10. Hanipah, M.; Mikalsen, R.; Roskilly, T. Recent commercial free-piston engine developments for automotive applications. *Appl. Therm. Eng.* **2013**, *75*, 493–503. [CrossRef]
11. McLanahan, J. *Barrel Aircraft Engines: Historical Anomaly or Stymied Innovation?* SAE Technical Paper 985597; SAE International: Pittsburgh, PA, USA, 1998. [CrossRef]
12. Axial Internal Combustion Engines. Available online: http://www.douglasself.com/MUSEUM/POWER/unusualICeng/axial-ICeng/axial-IC.htm#cd (accessed on 18 October 2020).
13. Chown, D.; Koszewnik, J.; MacKenzie, R.; Pfeifer, D.; Callahan, B.; Vittal, M.; Froelund, K. *Achieving Ultra-Low Oil Consumption in Opposed Piston Two-Stroke Engines*; SAE Technical Paper 2019-01-0068; SAE International: Pittsburgh, PA, USA, 2019. [CrossRef]
14. Mazuro, P.; Jasiński, D.; Teodorczyk, A. On the High IMEP Potential of Barrel Engines. In Proceedings of the PTNSS Congress, Radom, Poland, 16–17 June 2011; p. PTNSS-2011.

15. Heywood, J.B. *Internal Combustion Engine Fundamentals*; McGraw-Hill: New York, NY, USA, 1988.
16. Woodyard, D.; Latarche, M. *Pounder's Marine Diesel Engines and Gas Turbines*; Elsevier Science & Technology: Oxford, UK, 2009.
17. Asthana, S.; Bansal, S.; Jaggi, S.; Kumar, N. *A Comparative Study of Recent Advancements in the Field of Variable Compression Ratio Engine Technology*; SAE Technical Paper 2016-01-0669; SAE International: Pittsburgh, PA, USA, 2016. [CrossRef]
18. Tomazic, D.; Tomazic, D.; Kleeberg, H.; Bowyer, S.; Dohmen, J.; Wittek, K.; Haake, B. Two-Stage Variable Compression Ratio (VCR) System to Increase Efficiency in Gasoline Powertrains. In Proceedings of the DEER Conference, Dearborn, MI, USA, 16 October 2012.
19. Gowthaman, S.; Sathiyagnanam, A.P. Performance and Emission characteristics of Homogeneous charge compression ignition (HCCI) engine—A review. *Int. J. Ambient Energy* **2016**, *38*, 672–684. [CrossRef]
20. Dresner, T.; Barkan, P. *A Review of Variable Valve Timing Benefits and Modes of Operation*; SAE Technical Paper 891676; SAE International: Pittsburgh, PA, USA, 1989. [CrossRef]
21. Morton, R.; Riviere, R.; Geyer, S. *Understanding Limits to the Mechanical Efficiency of Opposed Piston Engines*; SAE Technical Paper 2017-01-1026; SAE International: Pittsburgh, PA, USA, 2017. [CrossRef]
22. Gao, J.; Tian, G.; Sorniotti, A.; Karci, A.; Palo, R. Review of thermal management of catalytic converters to decrease engine emissions during cold start and warm up. *Appl. Therm. Eng.* **2018**, *147*, 177–187. [CrossRef]
23. Achates Power: Industry Leading Engine Solutions. Available online: https://achatespower.com/ (accessed on 26 August 2020).
24. Khosla Ventures and Bill Gates Invest in opoc Engine Developer EcoMotors. Available online: https://www.greencarcongress.com/2010/07/opoc-20100712.html (accessed on 28 September 2020).
25. Herold, R.; Wahl, M.; Regner, G.; Lemke, J.; Foster, D.E. *Thermodynamic Benefits of Opposed-Piston Two-Stroke Engines*; SAE Technical Paper 2011-01-2216; SAE International: Pittsburgh, PA, USA, 2011. [CrossRef]
26. Pirault, J.-P.; Flint, M. *Opposed Piston Engines—Evolution, Use and Future Applications*; R-378; SAE International: Pittsburgh, PA, USA, 2010; ISBN 978-0-7680-1800-4.
27. Stroke-To-Bore Ratio: A Key to Engine Efficiency. Available online: https://achatespower.com/stroke-to-bore/ (accessed on 27 September 2020).
28. Kyrtatos, A.; Spahni, M.; Hensel, S.; Züger, R.; Sudwoj, G. The Development of the Modern Low-Speed Two-Stroke Marine Diesel Engine. In Proceedings of the CIMAC Congress, Helsinki, Finland, 6–10 June 2016; Available online: https://www.wingd.com/en/documents/general/papers/the-development-of-the-modern-2-stroke-marine-diesel-engine-(cimac).pdf/ (accessed on 28 September 2020).
29. Zhu, Y.; Savonen, C.; Johnson, N.; Amsden, A. *Three-Dimensional Computations of the Scavenging Process in an Opposed-Piston Engine*; SAE Technical Paper 941899; SAE International: Pittsburgh, PA, USA, 1994. [CrossRef]
30. Nieman, A. The Right Solution for Downsped Engines Dana Commercial Vehicle Technologies White Paper. Available online: https://leangreenfleet.com/documents/Dana%20Corp.%20downspeeding%20white%20paper.pdf (accessed on 28 September 2020).
31. Isenstadt, A.; German, J.; Dorobantu, M.; Boggs, D.; Watson, T. Downsized, Boosted Gasoline Engines. ICCT Working Paper. 2016. Available online: https://theicct.org/sites/default/files/publications/Downsized-boosted-gasoline-engines_working-paper_ICCT_27102016.pdf (accessed on 28 September 2020).
32. Lückert, P.; Schommers, J.; Werner, P.; Roth, T. The New Four-Cylinder Diesel Engine for the Mercedes-Benz B-class. *MTZ Worldw. eMag.* **2011**, *72*, 18–25. [CrossRef]
33. Mercedes-Benz OM 654. Available online: https://de.wikipedia.org/wiki/Mercedes-Benz_OM_654 (accessed on 27 September 2020).
34. Lückert, P.; Eder, T.; Kemmner, M.; Sass, H. OM 654—Launch of a New Engine Family by Mercedes-Benz. *MTZ Worldw.* **2016**, *77*, 60–67. [CrossRef]

MDPI
St. Alban-Anlage 66
4052 Basel
Switzerland
Tel. +41 61 683 77 34
Fax +41 61 302 89 18
www.mdpi.com

Energies Editorial Office
E-mail: energies@mdpi.com
www.mdpi.com/journal/energies

www.ingramcontent.com/pod-product-compliance
Lightning Source LLC
LaVergne TN
LVHW070930170726
843515LV00005B/912

* 9 7 8 3 0 3 6 5 1 4 4 8 2 *